How Information Warfare Shaped the Arab Spring

How Information Warfare Shaped the Arab Spring

The Politics of Narrative in Tunisia and Egypt

Nathaniel Greenberg

EDINBURGH
University Press

FOR MY PARENTS

Edinburgh University Press is one of the leading university presses in the UK. We publish academic books and journals in our selected subject areas across the humanities and social sciences, combining cutting-edge scholarship with high editorial and production values to produce academic works of lasting importance. For more information visit our website: edinburghuniversitypress.com

Edinburgh University Press Ltd
The Tun – Holyrood Road
12 (2f) Jackson's Entry
Edinburgh EH8 8PJ

Typeset in 11/15 Adobe Garamond by
Servis Filmsetting Ltd, Stockport, Cheshire

A CIP record for this book is available from the British Library
ISBN 978 1 4744 5396 7 (paperback)
ISBN 978 1 4744 5397 4 (webready PDF)
ISBN 978 1 4744 5398 1 (epub)

Contents

Figures and Tables

Figures

Tables

Acknowledgements

Work on this book effectively began on 25 January 2011 while my wife and I were living in the downtown Cairo neighbourhood of Lazoghly. On site to conduct research into the literary responses to the 1952 revolution, that opportunity was made possible by a grant from the Chester-Fritz Foundation of Seattle, Washington and the *Modern Language Quarterly*, then under the direction of Professor Marshall Brown. Since then I have received support or funding from the Center for Strategic Communication at Arizona State University, the College of Arts and Sciences at Northern Michigan University and the Mathy Junior Fellowship Foundation of George Mason University. Several conference and speaking invitations allowed me to gain critical feedback on aspects of the research. My thanks to Dr Sara Cobb and the Center for the Study of Narrative and Conflict Resolution at George Mason University, Dr Dina Matar and the Center for Global Media Communications at SOAS, Dr Marwan Kraidy and the Center for Advanced Research in Global Communications at the Annenberg School of the University of Pennsylvania and Dr Walid Mahdi and the Arabic Flagship Program at the University of Oklahoma. I am equally grateful to a number of individuals. In particular I should like to thank Dr Noha Mellor, Dr Naglaa Mahmoud Hussein, Dr Terri DeYoung, Dr Rei Berroa, Dr Israel Gershoni and Dr Jeffry R. Halverson, who allowed me to collaborate on several projects that proved progenitive to this one. Examples of our collaborative work that have resurfaced here include portions of our article 'Ideology as narrative: the mythic discourse of al-Qaeda in the Islamic Maghreb' (*Middle East Journal of Culture and Communication* 2017), and portions of chapter four from our book *Islamists of the Maghreb* (Routledge 2018). My thanks to the editors of *Jadaliyya* where portions of Chapter 3 appeared in the summer of 2015 and to the editors of the journal *Critique: Studies in Contemporary Fiction* that published portions of Chapter 8. More immediately I am grateful to the initiative of Nicola Ramsey of Edinburgh University Press who embraced this project unblinkingly from the start. Among those who provided insight into the ever-critical writing aspects of this book I should like to recognise Dr Kristina Olson of George Mason University, Dr Philippe-Joseph Salazar whose intellectual foresight became a major beacon for this project, and above all my father, Dr Edward S. Greenberg, who, unique among peers, has read virtually every word I have ever written.

Introduction:
The Hurricane and the Butterfly

Introduction

'*Le 17 décembre 2010, Mohammed Bouazizi s'immolait par le feu.*' 'On 17 December, Mohamed Bouazizi lit himself on fire.' The phrase would be repeated countless times in dozens of languages across the globe. I have copied it here from the book jacket to a short novel by the Moroccan author Tahar Ben Jelloun, *Par le feu* (By Fire), which appeared just a year after the events it relates. Ben Jelloun, a well-known novelist, was living in France when the uprisings began. Needless to say, he did not witness the incident in question. Few people did. Rather, he heard about it, as one does, gleaning the information from print or from sound, by word of mouth, or television broadcast. '*A young fruit vendor unable to make a living, a college graduate burns himself alive.*'[1] There was little reason to think this distant episode on a singular afternoon in a remote Tunisian city would transcend its instance of inception. But soon the story of Mohamed Bouazizi's death began to accumulate meaning, giving life and purpose to the otherwise arbitrary properties of its details. Soon the story was not his story. Nor was it about his death. In that story, the one told by those who knew him closely, Mohamed's death was the ending. For the rest of the world his death was the beginning, the 'spark' that ignites a revolution. That story was about illumination. And, in that story, Mohamed was not Mohamed. He was a metonym for despair and anger and a rising tide of discontent.

For an incalculable number of reasons when Mohamed's hand struck the match that lit his body afire, his life became part of a greater narrative, one that could be deployed and manipulated by novelists and politicians, activists and poets. The further one travelled from the epicentre of the incident the more important the narration of his life became. Adunis, the Arab world's most renowned poet, crystalised this phenomenon in a eulogy he delivered three months after Bouazizi's death in the industrial city of Sidi Bouzid where the event occurred: 'how did this Tunisian butterfly become a hurricane that would engulf the Arab world?' he wrote.

> The energy born by the event is harbored in the core of man. It ebbs and flows in spectacular fashion. It is not exclusive to the social elite, nor the upper class. It is not the product of some theory for the mass mobilization of society. Rather this energy has emerged from the lowest rings of society, from the lived experience of pain and suffering, from life itself.[2]

Such energy is magnified over space and time, powerfully and chaotically. Distinct from the refrain of 'domino effect' heard often over Western airwaves, the analogy of the 'butterfly' and the 'hurricane' invoked a less tangible link, an atomic impression gained on the ephemeral level of man's intuition.[3] This too is how narrative travels, over inter-waves and airwaves, through conversation and contemplation. Narrative, as such, is the antithesis of reality. It is that which signals the absence of the thing it describes, giving form and enumeration to something fluid and past. In this way narrative craves historical change. As Fredric Jameson observed in the wake of May '68, it has always been the case that with social upheaval 'history takes the form of narratable events, reveals itself as a continuity with a beginning, middle and end.'[4] By its very nature revolution upends the place of certainty and pulls into focus mankind's implacable desire for 'intelligible configuration'.[5] It is an indelibly fine line, however. The 'emplotment' of reality, in Aristotelian terms, is always at risk of admitting into focus the very 'heterogeneous collection . . . of intentions, causes, and contingencies' that compel it to operate.[6] In other words, as Paul Ricoeur wrote, just as narrative seeks to impose meaning on an otherwise incongruent field of information, it keeps in suspense – and retains the prospect – of precisely the opposite, namely, chaos.

Algeria, Sudan, Egypt, Bahrain, Yemen, Libya, Morocco, Jordan, Syria and Iraq. No corner of the Arab world was left unaffected. Following eighteen days of continued protest and the occupation of Tahrir Square, Egyptians witnessed their leader of thirty years step down. The ruler of Libya absconded in a blaze of violence. His final moments, bleeding and delirious, were broadcast around the world. Protesters in Jordan and Morocco staged days of 'rage'. The state of Bahrain saw demonstrations for over a month. Yemen and Syria collapsed into civil war. Libya divided. Criminal gangs proliferated across new autonomous zones, in the Sahel, the Sahara, Mont de Tebessa, the Sinai, the Gulf of Aden, and across the Levant. From within the storm powerful voices emerged. Parties formed. Elections were held. Despite the menace of terrorism, massive unemployment, unrest and economic collapse, Tunisia's democratically elected Constitutional Assembly succeeded in passing a new constitution. The politics of appeasement in Morocco, Jordan, Bahrain, Saudi Arabia, dimmed unrest. Much of what occurred remains a mystery, however, including for those in the countries involved. The following is a study not of the Arab Spring, *per se*, but of the mysteries that have surrounded it, of the 'poetry' that was the Arab Spring.

As I explore in this book, beginnings with no end in sight compel us to reconstruct linkages, to envision that which precedes and follows the perception of change. The uprisings were revolutions in part because they set in motion a proliferation of disparate information, creating new avenues of exchange and imagination. They unsettled presumed successions of power, redirected history and enhanced liberties – personal and collective. As spread by word of mouth, social media amplified the power of imagination. New vast and sprawling public spheres emerged online, on air and in the streets. Media makers, politicians, artists and activists jostled for prominence and, ultimately, authority. Yet, ironically, in the case of the Arab uprisings, the hyper-mediated sphere of twenty-first-century communications technology hastened the calcification of narrative just as it propelled into motion an unprecedented number of contributing voices. Narrative in this way weeds out imperfection and distinguishes dissent. It condenses fact into fiction and fiction into myth.

Ben Jelloun's 'Bouazizi' epitomised this phenomenon. The tale of *ibn al-suq* – a market boy – raised by the currency of community, antagonised by poverty, and poised, ultimately, to transcend it, extends to biblical times. The image embodies humility. It elicits empathy and anger towards the petty usurpers that would demand payment even from the most destitute of clients. It recalls a world of racketeers and thugs; a world left to its own devices in the shadow of a system that institutes lawlessness by way of intolerance, blind decadence and corruption. Ben Jelloun described his story as a fictional tale – *un récit*. For most, the meaning embodied by the name 'Bouazizi' meant something else, an idea less tangible, yet, strangely, more real. As Bruce Lincoln described the meaning of 'myth', the story of Bouazizi too was closer to reality than any artifice of fiction allows. It was 'ideology in narrative form'.[7]

Bouazizi, in his death, was flattened, delineated and multiplied like a paper doll. His story was to become timeless, asynchronous, wholly severed from the fear and doubt and flesh and blood that made him human. Such violence always underpins the narrativising of life – as French Hegelianists have long observed – deferring vital uncertainties to fixed formalities. The body becomes a contested space in such instances, quartered and drawn by a seemingly unending parade of those who would find substance in the remains of the flesh. In this way, an American president could reach across the ocean to ply the ghost of Bouazizi from the grip of his industrial heartland, his mourning family, lovers and friends, to prop him up in an alien land amidst the 'patriots in Boston who refused to pay taxes', or sitting 'courageously' like 'Rosa Parks'. Obama's speech on the Arab Spring seemed absurd on the surface: linking Bouazizi to this unorthodox genealogy of American heroes. But the illocutionary resonance of his message fitted within the context of his own interests. The 'patriots of Boston', a staple in Republican mythology, like the liberal hero of the Civil Rights movement, Rosa Parks, represented two ready-made poles in Obama's signature rhetoric of opposites. Bouazizi – who was, and remains, largely unknown by name to the American public – provided a grey area of reference to this discourse, drawing opposite poles of political reference into common orbit. As for the Arab cause, the reference to anti-colonialism and civil rights made sense too.[8]

The pliability of the message embodied by Bouazizi's ashes appeared limitless. Secular and sectarian explanations equally abstract in their appropriation of his name appeared almost instantaneously. A complex gendered discourse emerged to explain linkages between his death, his decision to die, and the events that followed. His name became code for new equations of power, justice and authority – none of which he knew in his lifetime. 'Belaid is not Bouazizi and I am not Ben Ali', explained the leader of Tunisia's Islamist party, Rached Ghannouchi, following the assassination of the country's labour leader, Chokri Belaïd, on 6 February 2013.[9] In defining the causality of events associated with his name, those who invoked it laid claim to prophecy, calibrating, through reason, that which seemed incalculable. The dissimilarity of a man on fire to the world around him naturally summons an artifice of explanation, firstly and secondly, rationalisation. But in doing so we look beyond the incident in question, delineating an imagined history, a series of causes and effects, both macro and micro. Of his forty-seven-page novel on Bouazizi, Ben Jelloun assented to the question, when asked, whether he had imagined in advance creating such a character.[10] The network of imagery and meaning surrounding the incident has been in place for centuries and it will remain in place long after Bouazizi's flame has dimmed.

On Verse and Verisimilitude

Captured in the digital age, the Arab uprisings epitomised the often transient notion that reality resists succinct narration. Or, at least, as I argue in this book, it tends to remain subject to a degree of *différance* in the Derridean sense. Over the course of the book, I look at how events on the ground in post-revolutionary Tunisia and elsewhere in the Arab world have not so much been 'deferred' by narrative, as they have been altered or reorganised to reinforce pre-existing ideologies. The experience of revolution, I suggest, offers a unique window onto the relationship between ideology, narrative and phenomena, simply insofar as the very definition of the term is predicated on the disruption of this linguistic troika. In simpler terms, it is no longer possible to determine exactly whose ideology is revolutionary and whose is not.

Narrative, I maintain, is predicated on the enumeration and compression of events into a diachronical sequence of action. Refined and reiterated through inquiry and consensus, the construction of narrative is a strictly human endeavour. And in the exploration of the new narrative patterns now framing the course of history in the Arab Middle East and North Africa, this book places equal emphasis on the ideological interests at stake in the reformulation of reality.

Witness Tunisia. On the eve of the revolution's third anniversary the Minister of Industry, Mehdi Jomaa, assumed the interim reigns of the Prime Minister's office. A so-called 'technocrat', Jomaa was charged with forming a caretaker government in advance of open elections to be held that autumn.

The announcement came in the wake of the Islamist party Ennahdha's sudden renunciation of the Prime Minister's office the previous September, ostensibly a sign of cooperation in the face of mounting criticism surrounding the government's failure to investigate the assassinations of two opposition political figures. A number of Western media outlets quickly absorbed the narrative advanced by Ghannouchi who referred to the appointment of Jomaa as a 'yielding of power'. This concessionary narrative, however, ignored the fact that neither of the Parliament's largest secular opposition parties supported the vote, nor had it achieved the requisite majority. Faced with mounting criticism, a spokesman for Ghannouchi's party, Ennahdha, denied reports in the French daily *Le Monde* from the previous day that the appointment had been directed by lobbying efforts from the US Department of State and the EU. In other words, Ennahdha's leaders defended the appointment as a victory as much as they sold it as a concession. The former lent itself to the long-standing critique on the part of secular pundits within Tunisia that Ennahdha had been playing a long game and was determined to alter the secular nature of the state. The latter suggested the Islamist party was representative of a democratic majority and envisioned a path of moderate conservative governance along the lines of the AKP in Turkey. Little more than an anecdote to the greater social convulsions that have gripped the region since the collapse of the old regimes, the truth of the actors' intentions may lie somewhere in between. But it is in examining the framework of the information that one finds a more rigid design, an ideological mandate of narrative that supersedes, by definition, any singular instance of history.

Existent solely in the absence of reality, narratives are inclined towards duplicity. There is no concrete explanation for why the story of Bouazizi travelled so far and generated as much interest as it did. It was a story conveyed through media in the digital age, but certainly it was not alone in this regard. It spoke to a range of issues ailing the developing world – youth unemployment most notably – but it was far from the first major story to do so. In Tunisia alone, a similar story surfaced just two years earlier – an unemployed youth from the country's industrial centre vanished and protests erupted – but few outside the country took notice. Multiple self-immolations occurred in the months prior to Bouazizi's as well but, as Marwan Kraidy observed, these remained 'resolutely local stories'.[11] No singular explanation could account for why this story, apart from all others, would elicit so much attention. From the daily barrage of news, seldom does any one story sustain the imagination of listeners beyond the moment of initial reception, having been 'shot through with explanation', as Walter Benjamin famously wrote.[12] In contrast, here was a story that appeared to have no ending. Volumes of commentary and scholarship on what led to the conflagration quickly developed. But each new day and every subsequent catastrophe seemed to extend yet further that which the fire in Sidi Bouzid set in motion. Is it surprising that the optimism infusing early discourse on the uprisings would soon give ground to an ever-encroaching cynicism? What rosy finish, after all, could be expected from an opening whistle muted by the roar of burning flesh? To flash forward then four years: another man burns alive, though this time in a cage and at the hands of a ruthless new force on the post-revolutionary stage. Can the rise of the so-called Islamic State be explained without reference to the Arab uprisings, and then, to that burning afternoon in Tunisia? Can it be understood dialectically, as a counter-revolutionary movement? Or indeed, perhaps, a culmination of sorts, begun not in Sidi Bouzid from a secular outcry against tyranny and poverty, but in nearby Kairouan as a holy struggle against decadence and apostasy? Each incident constitutes change, separate pieces of information welded together by narrative.

The simultaneously disruptive and constructive power of narrative is in effect everyday and its role is not reserved to the realms of media, politics or the arts. But it is also the case that these fields profit explicitly from what Hans Robert Jauss described as the 'horizon of expectations' in human communication,[13] that is, our desire to understand what we cannot see. In times of great social upheaval, narrative appears like a glowing pier in an ocean of collective darkness, lighting the way back to shore or, perhaps, further afield. Egypt's Tahrir Square embodied such a state of suspended disillusionment. Narrative, in all its duplicity, thrived in this space as it did across the greater global discourse on the Egyptian uprising.

Witness 'Khaled Said', a name and a face resurrected from the grave and propelled into a powerful digital afterlife on Facebook. The creator of Khaled Said, Wael Ghonim, was a young software engineer working for Google and living in the United Arab Emirates when he came across the image of the twenty-eight year old's battered face online. 'I was sitting in my small study

Figure I.1   Tahrir Square, 27 January 2011. © Nathaniel Greenberg

in Dubai, unable to control the tears flowing from my eyes', he wrote in his 2012 memoir.

> My wife came in to see what was wrong. When I showed her Khaled Said's picture, she was taken aback and asked me to stop looking at it. She left the room, and I continued to cry over the state of our nation and the widespread tyranny. For me, Khaled Said's image offered a terrible symbol of Egypt's condition.[14]

The image, juxtaposed to a portrait of the deceased while alive, became as much a point of departure for what lay ahead as it was a vestige of the past. Ghonim, at least, envisioned it this way when he used the page, along with the Facebook page of Mohamed ElBaradei (the former head of the International Atomic Energy Agency and founder of the National Association for Change), to promote the Egyptian uprising. 'The day they went and killed Khaled, I didn't stand by him', he wrote in his first post. 'Tomorrow they will come to kill me and you won't stand by me.'[15] Narrative was ingrained in Ghonim's use of Said's image from the outset. But like Bouazizi, the story it told eclipsed resolution leaving nothing beyond it but resentment or, perhaps, a willingness for revenge. Linda Herrera observed this point lucidly in her reading of Ghonim's first post on the Baradei page linking its followers to the new site. 'We invite all Egyptians to join the page of the Martyr Khaled Said who got tortured until his death by police in Alexandria', the post read. Its accompanying image was a traffic sign with the words: 'Point of No Return'.[16] In death Khaled Said's life was abstracted, his name and face folded unto one another like a diode through which the inconsistences of his existence would be channelled towards a singular if uncertain end.

The Nation Immemorial

While the machinations of narrative were hard at work in constructing the ideological framework of the uprising, counter-revolutionary currents were underway as well. As I discuss in the second chapter, amidst the convulsions of the first two weeks in Cairo the name 'Gamal Abdel Nasser' rose from the past to summon order. An outsider could be forgiven for not comprehending the popular reception of the military's arrival in Tahrir as one of deliverance. The 'Police Day' demonstrations sparked by Khaled Said and other like causes online ostensibly targeted the regime's *modus operandi* of martial law. Yet the invocation of 'Nasser', the leader of the 1952 Free Officers' Coup, also bespoke revolution.

Promulgating this powerful and ultimately victorious narrative was Egypt's standard-bearer of the old lettered class. *Al-Ahram* newspaper is the oldest daily in Egypt. Beginning in 1875 as a chronicle of market life and economics in Alexandria, the paper has long served as a beacon of Egyptian thought.[17] But since its nationalisation in 1961, the paper has provided more of a vanguard role than a responsive one in the formulation of the government's ideology, a tradition started, above all, by Mohamed Hassanein Heikal who assumed the Chief Editor position in 1957. As Nancy B. Turck reported for Saudi *Aramco World* in 1972, 'the "authoritative" part of the cliché ['the authoritative *Al-Ahram*'] stems largely from the unshakable conviction of readers that Chief Editor Mohamed Hassanein Heikal, former friend and confidant of the late President Gamal Abd al-Nasser, is the semi-official voice of the Egyptian Government'.[18]

Though no longer the Editor-in-Chief of *Al-Ahram*, Heikal would emerge once more in 2011 to concentrate the bluster of Tahrir into a clear and definitive narrative that eclipsed virtually all other positions and helped to set in motion the revolution's ultimate resolution. After a series of profound developments on the ground, Heikal appeared, first on the covers of the nation's leading opposition papers and later *Al-Ahram* itself, expounding a description of events that recast the occupation of Tahrir from a social-media fuelled movement for democracy to a military-led deliverance from chaos, firstly, and a Western-backed conspiracy, secondly. Despite the attention from global media, none of the online voices or pro-democracy leaders in the street could articulate a narrative that resonated as clearly with Egyptian history as Heikal's. And while the inference of a military vanguard in the service of revolution carried historical roots, the discursive vehicle for ushering the idea forward in the context of Tahrir had a shallower past.

For a month leading up to 25 January, terrorism had dominated public discussion. The Two Saints Church attack in Alexandria on New Year's Day was the country's deadliest terrorist attack since the Luxor massacre in 1997. With little headway on the investigation, rumours about the bombing were rampant. This had not stopped Egypt's state-sponsored media industry from launching a multifaceted reconciliation campaign. Politicians and celebrities used the 24 December bombing to promote national unity, pride and dignity. Government officials staged a series of high-profile appearances at the church. Celebrities and politicians visited Coptic neighbourhoods and churches. For the first time in modern history, the President, along with the head of al-Azhar, Africa's oldest centre of Islamic learning, attended Coptic Christmas ceremonies on 7 January. The occasion was broadcast on national television – also a first. On 7 January, a state-commissioned music video featuring some of the country's most renowned stars was recorded on the steps of Cairo University.[19] Thematically, the song, '*Masr muftah al-haya*', was an exercise in nationalism. Composed, written and directed by Amr Mostafa, Medhat al-Adil and Mohammed Bekir – each a major figure in Egypt's heavily regulated film and television industries – the operetta featured stars Mohammed Hunaydi, Mohammed Fu'ad, Sharif Munir, and others dressed in black and foregrounding a dimly lit choir. 'No and a thousand nos!,' the chorus followed. Bold, even melodramatic, the performance illustrated a clear departure from the state's relative silence following previous attacks.

But here, as with the other string of cultural productions to appear in the first two weeks of 2011, the attack in Alexandria became a mere phantasmal backdrop to the performance of national solidarity. Nowhere was this effect more prominent than in Hani Shaker's minor blockbuster '*Eid wahda*' ('All Hands Are One').

'*Kol muslim aw masihi fiha 'aysh wa bi-karama*' – 'All Muslims and Christians live in dignity' goes Shaker's song. A polished, almost industrial kind of national ode, '*Eid wahda*' like '*Misr muftah al-haya*' epitomised a mode of cultural identity that had been resurgent in the country for over a decade. Oriented around themes of tolerance and pluralism, Shaker's lyrics embodied an older kind of nationalism, a 'Mediterranean' value system that was free of any evident allusion to Pan-Arab socialism,[20] but also, at the same time, deeply apolitical and closely tied to the corporatist structure that had bolstered Egypt's military regime from the time of the Free Officers' Coup. Less a philosophical repudiation of intolerance in favour of pluralism, '*Eid wahda*' evinced an imagined social equilibrium, a neutral space born in contradistinction to the monopolisation of faith by extra-official Islamist voices (i.e. all Egyptians, not just the Brotherhood possess 'dignity') on one hand and, on the other, a less certain, though invariably foreign element of instability. It was this dimension of the axiom Heikal reintegrated into his discourse on the revolution that ultimately gained traction in the form of popular song and sloganeering.

In truth, of course, the demonstrations had called for much more: an end to corruption and misery ('Bread, Justice, Dignity'). The Square was marked by the eclecticism of its occupants' ambitions, a diversity that reflected the reality of the Egyptian street. Intertwined with this fluidity were emergent cords of consensus but these too were myriad. How to name what was occurring? Was it the January 25 Revolution? – the end product of an 'electronic enlightenment' as writer Mahmud Uthman had prophesised in his 2009 novel *Revolution 2053*. Or 'The Revolution of Rage'? – as the opening day of protest (*Youm al-Ghadab*/Day of Rage) was initially advertised. 'The Revolt on the Nile' (*Thawrat al-Nil*) was one description that emerged early on. 'The Camel Revolution' (*Thawrat al-Jimal*), in reference to the Battle of the Camels, on 2 February, another. The description 'Lotus Revolution' (*Thawrat al-Lutis*) faded more quickly than some thought it would. As

Mekkawi Said wrote in his 2014 *Kurrasat al-Tahrir* (The Tahrir Papers), the ancient undertones of the struggle – the walls of the city adorned in twenty-first century hieroglyphics, and the earth-shattering prospect of a Pharaoh's fall – were captured elegantly by the image of the Nile flower. More famously still at the time was the proclamation of then Secretary General of the Arab League, the Egyptian Amr Moussa, that the 'leaderless' movement was a 'White Revolution' (*Thawra Bayda*'), absent the stain of blood.[21] Alluding to the so-called 'White Revolution' of the Pahlavi regime in Iran, the title was an ominous one. As with the 'Shah's Revolution', Egypt's military would emerge as saviours, newly empowered by an ostensibly popular cause.

The semantics of the revolution adorned events like a string of pearls worn easily by one statesman or another. Before Musa's observation, it was the military who called on the 'great people of Egypt' to protect themselves from 'terrorists' and to preserve the interests of the nation from would-be looters.[22] The manifestation of 'white weapons' (*aslaha bayda*') precipitated his description of *Thawra bayda*' (White Revolution). As Mekkawi Said recalled humorously the head of al-Azhar warned in advance of the Friday of Rage that protests would send the country to the 'brink of collapse' (*haffat al-inhiyar*). After the revolution it was the 'brink of a new stage' (*haffat marhala jadida*) in the country's social and political history.[23]

Conclusion: Narrative and Revolution: the First Three Years

In the following seven chapters I explore some of the many myriad angles of revolution as an expression of form. This includes inquiry into the viability of narratology as a method of social critique. While an imperfect science, I argue it is now more important than ever to rethink how narrative shapes perception. Millions of dollars have flowed into this equation as counter-communications experts and politicians seek to gain footing in what Evan Osnos, David Remnick and Joshua Yaffa have described as the 'New Cold War'.[24] But as I show in Chapters 1 and 2, much of the information warfare we see today is recyclical. As an instance of radical collapse and rapid recovery, the informational landscape of the Arab Spring offered a paradigmatic glimpse of a far broader proxy communications struggle. In Chapter 2, along with examining the framing of the first eighteen days by two iconic Egyptian newspapers, I examine how the influence of WikiLeaks as well as the US programme of Cyber Dissident Diplomacy (CDD) trickled into the main-stream of domestic public discourse. While events on the ground evolved in breathtaking and at times wholly incoherent fashion, the fastidiousness of narrative served to crystalise the inherent conceit of the myriad communications operatives seeking to leverage the future against the present and the present against the past.

This was particularly vivid in the immediate post-revolutionary media sphere of the Arab Spring capitals. As I show in Chapters 3, 5 and 6, the discourse on democratisation in Tunisia and Egypt quickly gave way to more time-honoured stories of unification, securitisation and, in the case of Tunisia especially, *ijtihad*: a quite literal description of social critique.

In Chapter 3, I also examine how the bizarre rise and fall of Tunisia's 'Public Enemy Number One', Abu Ayadh, appeared emblematic of what Plato famously described as the 'paradox of demoracy': that that which democracy defines as 'good', namely 'freedom', is also what destroys it.[25] While Abu Ayadh's radical trajectory careened towards disaster – a mysterious disappearance in Derna, Libya – Tunisian media used his name and the danger associated with it to regain hold of the centre. A similar pattern can be found in the peculiar ebb and flow of censorship laws, the history of which I examine in both the case of Egypt and Tunisia, in Chapters 5 and 6, respectively. The collapse of the old regimes in 2011 enabled an unprecedented window for creative and political expression. But it also set in motion a new array of ideological alliances, what I describe in borrowing from the work of Paul Ricoeur as 'narrative identities'.[26] Colliding against the transformative events of 11 September 2012, post-revolutionary politics in both countries reassembled along new ideological fault lines, all of which were defined by a reconceptualisation and narrative formulation of the uprisings' past.

The rhetorical recistribution of power in the form of censorship was particularly evident in the domestic press. In August 2012, after an attempt on the part of government officials from Ennahdha to replace the head of Tunisia's largest publishing house, Dar al-Sabah, with one of their own party members, the newspapers *Le Temps* and its Arabic counterpart *Al-Sabah* went dark. In his column from 25 August, the last for nearly four months, Youssef Seddik, one of the country's most renowned public intellectuals, wrote of Diogenes and the myth of the Cynic carrying a candle in the daytime as he searched for an honest man. 'If you allow me to contribute my modest stone to the construction of this debate', he wrote, 'why not suspend this wicked decision and allow time for an appointed commission of tradesmen to find the best and most creative candidate and the one most likely to improve circulation.' Like the petition filed by the heads of *Al-Sabah*, his plea a la 'petit Diogène', was addressed to the 'three Presidents' (Moncef Marzouki, Hamadi Jebali and Mustafa Ben Jaffar). But the real addressee, as with many of his columns, was Rached Ghannouchi: *'celui par qui tous les scandales arrivent ou n'arrivent pas, s'il le veut bien'* ('he to whom all scandals arrive whether he likes it or not').[27] 'You have the chance', he wrote, to become the 'honest man' for whom the Cynic searches. Efforts to repeal the appointment failed. Employees staged a fifty-seven-day strike, including a nineteen-day hunger strike. And the crisis subsided only after representatives of the UGTT, the country's powerful labour union, succeeded in negotiating a resolution.

As Seddik understood, the incident signalled a turning point in relations between the recently elected Ennahdha leadership and Tunisia's secular punditry class. His daily column, known simply as *La chronique de Youssef Seddik* would become, between 2011 and 2013, a beacon of critical reflection on the otherwise unfathomable flow of events. As I discuss in Chapter 6, here too narrative became the philosopher's indispensable counterweight, not least in the degree to which he fashioned his description of events around an indispensable, if implacable pole of gravity, namely, Rached Ghannouchi: the philosopher's foil.

Chapter 7 swings to the opposite side of the ideological spectrum in the Maghreb to examine how narrative was used to frame events by one of the region's most dangerous, but increasingly influential ideological consortiums: the organisation known commonly as al-Qaeda in the Islamic Maghreb (AQIM). Drawing on a broad sample of the group's internal and external communiqués, this chapter provides for the first time a comprehensive analysis of AQIM's discourse on the Arab Spring. Not surprisingly, AQIM's protagonists and antagonists, its imagined beginnings and foretold ends, are radically different from most mainstream narratives of the uprisings. Yet, common across much of their rhetoric was a shared emphasis on securitisation.

In the communiques of AQIM, the promise of absolute security in the form of sharia emerges as the veritable beginning and end of the revolution's arc. The ideological, as well as personnel, crossover between this organisation and the Islamic State group (*al-tanzim al-dawla*) was all but seamless. But was theirs a revolutionary narrative or a counter-revolutionary one? And if so, how does the narrative of AQIM and the Islamic State emerge in contrast to other ostensible 'counter-revolutions', most notably, the rise of militant nationalism? As I discuss in Chapter 7, many of these questions fall beyond the realm of reason. But in the age of digital reproduction, the 'lifeworld context' of communication precipitates a mode of rationalisation that expects the absence of reality.[28] In the age of digital reproduction, I argue, authenticity is irrelevant and myths make sense, just as much as the truth, if not more.

In this way, art – as has always been the case – must be understood as the primordial scene of narrative dialectics. It is the mental and emotional space in which reason gives way to impression and impression to empathy. In the realm of art, some of the most striking symbolic reversals of the Arab Spring gain a renewed sense of historical meaning: from the cinematic afterlife of the feared *baltagiya*, the veritable *agent provocateur* of Tahrir Square, to the rise of speculative fiction with its richly imagined dystopias of the revolutionary present. For this reason, the book concludes with a discussion of narrative in its 'natural habitat'. With context in place, it is here, I suggest, that future generations will most likely turn as they seek guidance from the past for the revolutions to come.

I

Information Warfare 2.0:
A Methodological Critique

Introduction

In a 2008 study, *The Science of Stories*, Hungarian social-psychologist János László recounts a provocative experiment from the annals of Gestalt psychology. In the early fifties, Solomon Asch, then a Professor at Swarthmore College in New York, conducted a series of perceptual tests aimed at better understanding 'group effects on judgements and attitudes'. Using students from his classes, Asch presented groups with two white cardboards, one displaying three vertical black lines of varying height and the other a single black line. Asch then asked the group to say which one of the three lines was the same length as the standard single line. He found that when students submitted answers privately in writing, their responses were mostly accurate. In one group of twenty-five students, just 7.4 per cent submitted erroneous answers. The accuracy of the students' judgement changed dramatically, however, with the implementation of an additional variable, namely, majority opinion. Asch demonstrated this by running another set of experiments that entailed having the majority of students in the class, all but a small handful, secretly agree to provide an erroneous answer for the matching lines exercise. Members of the 'cooperating' group, as Asch wrote, were instructed prior to class 'to act in a natural, confident way, to give the impression that they were new to the experiment, and to present a united front in defending their judgments when necessary'.[1]

The result of the experiment run with multiple groups was that one-third of all non-cooperating students were persuaded to agree with the majority. Or as Asch wrote: 'the erroneous announcements' of the conspiring majority 'contaminated one-third of the estimates of the critical subjects'.[2]

Asch's experiments demonstrated that the views and opinions of people can be powerfully shaped by the influence of those around them even when there is tangible evidence to the contrary. This observation could hardly be more relevant today, a time in which disinformation is being used intentionally by governments and non-governmental actors to bend popular narratives in ways that have material effects in the real world. Most evident at the time of writing was the disinformation campaign that helped Donald J. Trump win the American presidency in 2016. The engineering of falsehoods – or 'misinformation' – that contributed to his election did not begin with Trump's presidential campaign, of course. Rather, it could be said, by 2016, the phenomenon was endemic and his election was the result of a contagion that had spread the world over. So-called 'merchants of doubt',[3] 'ideologically motivated' and interest-vested advocates of misinformation have become prolific in the Internet age, swaying hearts and minds over any number of hot-button issues.[4] The problem is so pronounced that one group of scientists at Cambridge University studying misinformation campaigns around climate change proposed creating a 'fake news "vaccine"', pre-emptive doses of untruths so that people may develop 'a cognitive repertoire' to 'build up resistance to misinformation'.[5] János László identifies another mode of resistance to the persuasive power of collective falsehoods. Psychologists revisiting 'the Asch situation'[6] in the mid-1970s found that when subjects attributed 'rational motives' to the 'incorrect judgements of their peers' the subjects were able to effectively dismiss the collective lie and to provide 'non-conformist correct responses'.[7] In imagining what 'rational motives' those cooperating in the lie might share, the subject is able to piece together a kind of history, to assign those cooperating in the lie a place within the *dramatis personae* of the performance. He is able, in other words, to determine a narrative.

Of course, no such control as 'Asch's situation' exists in the real world where we are bombarded by information, true and false, at an incalculable rate. Nor is there any prospect of an actual 'vaccine' to counter the latter. Yet as certain historical landmarks in this new century have shown, the proliferation of uncertain information, while blinding in its glare, has not diminished our will to ascribe coherent intention, or motivation, to the dissemination of information.

The so-called Arab Spring was one such landmark. The history recounted in this book happened in the midst of a technological revolution. It is known, for example, that mobile phone usage in Egypt had reached a penetration rate of 100 per cent by 2011, up 50 per cent from just three years earlier;[8] that the number of Internet users in Tunisia nearly tripled from 17 per cent in 2007 to 39 per cent by 2011;[9] that satellite news channels, including *Al-Jazeera*, *Al-Arabiya*, *France 24*, *BBC Arabic*, *CNN Arabic*, *RT Arabic* and many more had reached nearly every corner of the Arab world. The uprisings greeted the average spectator with a veritable deluge of information. But amidst this maelstrom and, indeed, because of it, bold and singular narratives emerged to ascribe meaning, to define intentions, and to elucidate the subjective dimensions of reported reality.

The Perception of Narrative/Narrative Perspective

The mode of narrative analysis I pursue in this book is one of rolling back such ascriptions. While the assigning of rational action to dubious events signals an attempt on the part of the narrating subject to connect 'the what and the how to the who',[10] analysis of this coda threatens such 'hegemonic' ties.[11] The 'what', the 'how' and the 'who' become part of a common schema, the value and meaning of which is bound to the 'lifeworld' context of the narrator.[12] This method of analysis is peculiar to the discipline of literary studies insofar as it posits that the primary charge of the critic is to distinguish 'art as experience'.[13] But it is also particular to the primary source material in question. I have used previously the phrase 'emergent public discourse' to articulate the range of discursive ephemera employed in this research: social media posts, newspaper articles, talk show discussions, pop songs, political speeches, protest slogans, as well as essays, novels, memoirs, poems and interviews.[14] Across this body of grey material, the book seeks out patterns – recurrent rational attributions, artificial beginnings, and desired ends.

Here the bulk of narrative in question emerged concomitantly with the experience of the narrator. That is, 'the narrating person and the narrative event are located in the "here and now".' As described by László this particular kind of 'narrative perspective' is closely interwoven with the narrator's 'identity state', that is, the static reservoir of personal clarity through which a rationalisation of the kinetic world flows. Narratives produced contemporaneously with the narrator's perspective often contain 'unresolved identity-conflicts' because the narrator's capacity to rationalise and contextualise surrounding contingencies appears in disarray.[15] The second narrative perspective, in which the narrator is located in the 'here and now' but the narrated event is located in the 'there and then', is most often a device for absorbing such identity-conflict. This perspective is most often the voice of the statesman, the journalist, the lawyer and the academic. Here we find heroes and victims, losers and winners. The narrative perspective of 'there and then' shows us cause and effect, a backstory and, at times, an ending. The narrative perspective of 'there and then' contains crisis within the context of its own design.

The contrast between these two narrative perspectives was on sharp display amidst the opening salvo of unrest in Egypt. For the independently owned and oppositional newspaper *Al-Shorouk*, the opening events of the uprising fitted neatly into a pre-existent narrative of social mobilisation and civil disobedience. In contrast, *Al-Ahram*, the state flagship paper, struggled at first to provide a coherent narrative of the unfolding events, electing at first to avoid coverage of the protests all together. As events progressed and the military regained control, however, *Al-Ahram*'s coverage gravitated towards the 'there and then'. Following the lead of Mohamed Hassenein Heikal, the country's pre-eminent public intellectual, the now familiar narrative of military-led liberation began to emerge on the front pages of *Al-Ahram* concomitantly with the newspaper's positive identification with the occupation of Tahrir. Meanwhile as the 'here and now' of events on the ground overwhelmed the coherency of *Al-Shorouk*'s opening line of reporting, the liberal opposition paper drifted into a more 'intense' field of discourse as it began what would become a long, precipitous descent into a state of 'identity-conflict'.[16]

The onset of reporting surrounding Mohamed Bouazizi's self-immolation in Sidi Bouzid presented a similar display of conflicting narrative perspectives. While reportedly discovered through a Facebook post by researchers at the new media desk of *Al-Jazeera*'s Doha bureau, Bouazizi's act was almost instantly filtered through the lens of 'there and then'. Within a matter days the unemployed fruit vendor had been cast as a symbol; his act of immolation a sign of indignation in the face of a cruel and repressive regime.

On the surface, such narrative coherency was ironic in light of the incident's apparent spontaneity as well as the manner in which it was first reported, namely: the eye-witness accounts of bloggers and activists 'on the ground'. Yet the rapid conversion of the Bouazizi incident into the Bouazizi myth was not so much remarkable for its artifice (we know now that many of the early details surrounding his identity and his immolation were apocryphal), as it was for the unanimity through which the narrative was affirmed as reality.

In general, an eerie sort of *sensis communus* formed around the edges of the initial uprisings. In Cairo, the effect appeared all the more peculiar to me once access to international media became available again. As I wrote at the time, the actual drivers and objectives of the sit-in were anything but obvious to me and others around me then living in Egypt.[17] In conversation with an anonymous Egyptian security official in the closed chamber of his office just beside the Ministry of Interior, I recalled for him, on the eve of the Million Man March, the occupation of the Zócalo in Mexico City by supporters of Manuel López Obrador following his narrow presidential defeat in 2006. Supporters had been allowed to protest peacefully and within several days dispersed without incident. He was amused and we both hoped for a similar outcome. A certain spirit of the Zócalo prevailed in Tahrir, but, of course, as history has shown, the Egyptians were bound for a fully different course. There appeared in fact no precedent for the eighteen-day shutdown of Africa's second largest city. Yet to watch the scene unfold on *CNN* or *Al-Jazeera* or any number of blogs, e-zines or news outlets around the world, was to witness history. The reproduction of the celebratory masses, clipped and replayed *ad infinitum* at once added to and detracted from the very reality depicted therein. The uprisings, to quote Gadamer, were subsumed beneath an 'ocean of the beautiful', an aestheticised vision of liberal contestation seemingly vast enough to encompass the world.[18]

Narrative Crowd Sourcing

But where does the techno-aesthetic effect of the 24-news cycle end and strategic communications begin? In the case of the Egyptian uprising one avenue for addressing this question can be found in examining the elaborate, if largely misunderstood work of WikiLeaks.

Beginning on 27 January, Julian Assange and his organisation released a string of documents from the cache of State Department cables stolen the previous year. Different documents were delivered to different news agencies for different purposes. At *CNN*, which published its report on the morning of 28 January with the headline 'U.S. cables: Mubarak still a vital ally', editors received a document in which State Department officials are seen expressing support for Mubarak, one of their strongest allies in the region, despite 'frustration with Mubarak's reluctance to address human rights issues' and 'a growing unease with the lack of a succession plan'.[19] The information legitmised fears of a dynastic takeover while deepening the wedge between the regime and the people by further confirming the dictator's reliance on Washington. *The New York Times* reported different information. Here we learn, on the night of 27 January, that 'relations with Mubarak' had 'warmed up because Obama played down the "name and shame" approach of the Bush administration'. The memoranda left 'little doubt about how valuable an ally Mr. Mubarak has been' and indicated that the State Department was 'pessimistic' about the 'prospects for the April 6 movement'. The *Times* highlighted the ongoing struggle between the Egyptian government and American-backed Non-Governmental Organizations (NGOs) in Egypt and noted that the US Ambassador to Egypt Margaret Scobey had asked that 'three American pro-democracy groups be granted formal permission to operate in the country'.[20] When I spoke to Margaret Scobey, she clarified that the American position was never one of advocacy for NGOs like the National Democratic Institute, the International Republican Institute, or Freedom House *per se*, but rather, simply, clarity. The existent Egyptian policy in 2009 obliged such groups to operate in an unofficial manner, neither sanctioned nor prohibited. The policy, which was typically associated with the Minister of International Cooperation Fayza Abul-Naga, was a strategic one on the part of the regime, an explicit effort to keep in play the notion that the NGOs' pro-democracy work was clandestine and thereby suspicious.[21] By creating the façade of an archival document, however, the WikiLeaks disclosure to the *Times* allowed for no such nuance. With the imminent surge of 28 January on the horizon, the leaked cable pertaining to the subject of NGOs effectively weaponised the narrative of US interference, not least because America's paper of record was covering it as news.

Suspicion of collaboration between Washington and pro-democracy actors was made more vivid by a simultaneous cache of cables that appeared at this time in European newspapers. On 27 January, the *Aftenpost* in Oslo, and *The Daily Telegraph* in London on 28 January, published a set of stolen memos that told a wholly different story of US involvement in Egyptian politics. The European sample, as *The Telegraph* exclaimed in its headline, included information about 'America's secret backing for rebel leaders behind [the] uprising'. Here we learn how the American Embassy in Cairo had facilitated covert training for a member of the April 6 group, how in 2009 the individual met with House, Senate and Think Tank staffers in Washington, and how, according to Ambassador Scobey, 'several opposition forces agreed to support an unwritten plan for a transition to a parliamentary democracy, involving a weakened presidency and an empowered prime minister and parliament, before the scheduled 2011 presidential elections'. Along with the story, *The Telegraph* published a link to the 'confidential' memo hacked from the US Embassy.[22] The article generated 1,382 comments and 23,000 shares on Facebook. The comments section, a known territory for internet trolls,[23] was rife with rarified hostility. 'Here is the truth about Wael Ghonim and his partners claiming that what happened in Egypt is a spontaneous youth revolution', writes one commentator in Arabic and English. Another writes in Arabic: 'Look what these writers are talking about . . . the documents from WikiLeaks confirm the role of the CIA in everything that happens.' Another user identified as 'Tropicgirl' evinces a fully-fledged artifice of narrative conceit (see Figure 1.1).[24]

As with the hacking of the American elections in 2016 by Kremlin-linked cyber-espionage agencies, the opening of the Arab Spring set in motion a complex interchange of counter-communications operations and agendas. The 'wolves', as it were, 'were everywhere'.[25]

'Tropicgirl', who would become active across a broad if peculiar range of media including *The Daily Telegraph*, *The Hill* in Washington, *Breitbart News* and *Investment Watch Blog*, was responsible for 57,400 comments as of March 2017, far more than any single human being could produce. Imbued with the kind of racially-pinged conspiracy jargon used to influence the American election in 2016, the reliably ardent 'Trumpist' (as the user later declared), gave casual evidence to a near fully-formed messaging strategy that, by 28 January 2011, was deftly poised to descend onto Egypt's emerging

> tropicgirl · 7 years ago
> Wow... Looks like OBanana and the New World Order are actually throwing Israel under the bus, along with any other quasi-democratic Arab state.
>
> Wow... And its not just a threat, it is now actually happening. Probably already done.
>
> Well, I guess this just shows that Israel and the globalists are not "one". Everyone that was blaming Israel for a l sorts of things, not that the leaders did not act in an evil manner, were not quite on the money. The thieving world bankers are, apparently not pro-Israel, or, certainly not pro Israeli people.
>
> So, I guess everyone played nto the globalists' hands. Israeli thugs, Arab thugs, renditioners, torturers, flotilla blockers, and the rest of the faithful, all thrown off the cliff, used and abused.
>
> Wow...Like making that deal with Iran (Contra) while they were holding our people. Looks like the New World Order would rather have a Stone-Age Caliphate than democracies. After all, where there is forced starvation, like in Egypt, you can always smell the work of the globalists.
>
> Wow... I really did not expect this...
>
> The only hope is to stoke the new Arab governments toward representative democracy, (AND ISRAEL, for that matter, or whatever will be left of it) as much as possible and counsel them, JUST LIKE THE TEA PARTY... DON'T MAKE ANY CROOKED DEALS WITH THE STATUS QUO BAD ACTORS, or in their case, NEW WORLD ORDER (CFR, Britain, US, Goldman Suks, etc). Learn a lesson. And again. God help these poor people with wolves everywhere.

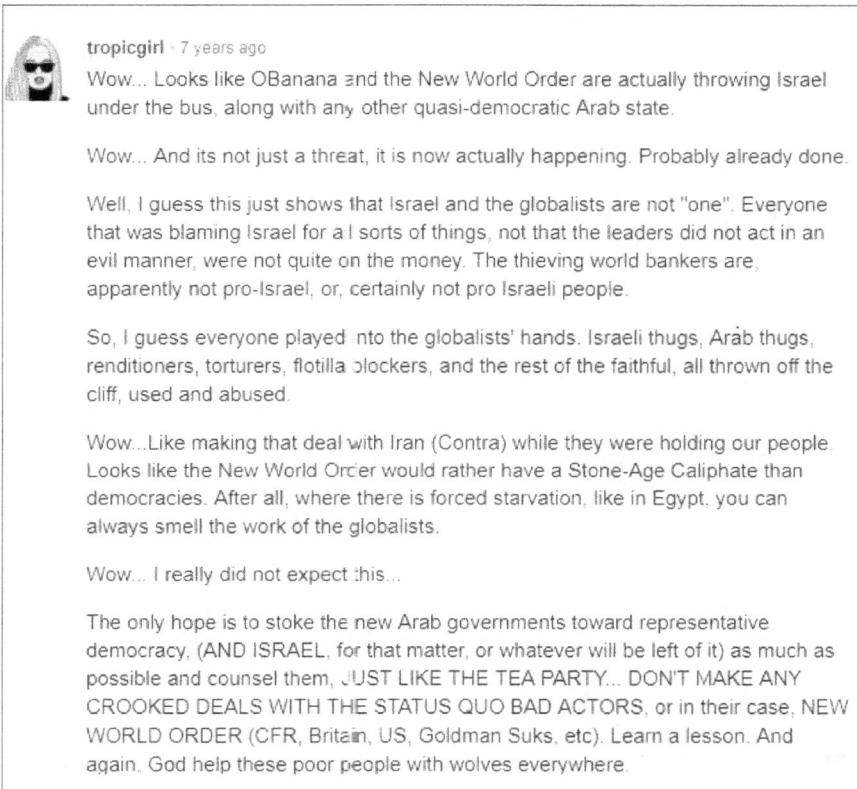

Figure 1.1   Digital troll fuels the counter-narrative, 28 January 2011

field of public discourse. But the careful calibration of such trolling, artfully drafted yet steeped in a jarringly studied discourse of indignation ('globalists' – 'world bankers' – 'New World Order'), also revealed a thoroughly rational chain of resistance.

Russia's active defence of the Assad regime in Syria was apparent. Its cyber offensive into the political turmoil of Egypt, America's largest ally in the region, was stealthier. With the Internet shutdown, a response, the government claimed to a massive Distributed Denial of Service (DDoS) attack launched against the country the day prior, *Al-Ahram* published a translation of the *Aftenpost* article on 29 January. Whether driving the narrative or supporting it, the WikiLeaks exposures contributed forcefully to the demonisation of the 25 January activists and their preferred medium of communication by Facebook – what Julian Assange called an 'appalling spy machine'.[26] Though not exclusively the cause, democratic activism in Egypt has yet to recover from the aura of conspiracy, secret US dealings, and the fog of cyberwarfare. The country's relationship with Russia, however, has never been stronger.

War of Ideas

In the realm of strategic communications, Russian and WikiLeaks effort to covertly shape the narrative of the Arab Spring through the use of foreign media was not unprecedented. To the contrary, outsourcing influence has long been understood as an effective method for multiplying the number of voices engaged in advancing a given cause while also normalising the work of propaganda through the perception of independence.[27] In 2007, the US National Strategy for Public Diplomacy and Strategic Communication described the importance of utilising 'key influencers' for engaging 'vulnerable populations' (youth, women and girls, and minorities), and of reaching 'mass audiences' through 'foreign media'.[28] Public Diplomacy 2.0, a State Department initiative designed to encourage 'young people to follow productive paths' by creating a cultural environment that is 'hostile to violent extremism', enlisted a host of partners from a range of corporate and civilian non-state actors to convey the message. 'Expertise resides in the private-sector', exclaimed James K. Glassman, the Under Secretary for Public Diplomacy and Public Affairs and the chief architect of Public Diplomacy 2.0 (PD 2.0). 'Our job is to find it and use it and to serve as a partner.'[29] In a 2008 speech at the New American Foundation, Glassman described one such early collaboration. Working with companies and organisations as diverse as NBC Universal, the Directors Guild of America and the Tisch School at NYU, the State Department created an online 'Democracy Video Contest'.

> Entrants make their own three-minute videos, posted to a site on YouTube, with the topic, 'Democracy Is . . .' Winners will be determined by a vote of the public over the Internet. While we did set a few rules – no pro-terrorist or pornographic videos – it is certainly possible that the winner of the contest will espouse views not completely shared by the U.S. Government.

Organize a smart mob—a group of people who mobilize on short notice to perform a collective

Figure 1.2   How to organise a smart mob. Howcast Media, YouTube, 2009

Elsewhere the government worked with major marketing firms to craft more content-specific campaigns. Howcast Media, for example, an early collaborator in the State Department's Alliance for Youth Summit of 2008 (a by-product of PD 2.0), created a series of videos designed for promoting, in Linda Herrera's words, 'human rights blogging, civil disobedience, and social media campaigning'.[30] Appearing on YouTube in 2009, the videos featured a multi-ethnic cast of well-dressed young actors performing neatly scripted roles in front of a green screen. The contrite, didactic dialogue is unabashedly propagandistic. 'Did you know in 2001, Filipinos responded to forwarded text messages by gathering in a Manila plaza?' explains the narrator of the two-minute video 'How to Smart Mob', 'The demonstration helped bring down the government of Joseph Estrada.'[31] As Herrera observes, the videos were less remarkable for their political edge than their implicit emphasis on branding and monetising an already existent political will.[32] Still, a number of the tactical components promoted by the collaborative messaging strategy undoubtedly gained traction among certain segments of the target population.

Not surprisingly, many of the products created as part of the Public Diplomacy 2.0 initiative remained closely tied to their aesthetic origins as online marketing ploys. And much like advertisers who fashion their

How to Smart Mob

12,705 views

Success!

Figure 1.3 'How to Smart Mob'. Howcast Media, YouTube, 2009

campaigns to disrupt cognitive association between the products they are selling and those seeking profit from their purchase, PD 2.0 sought to divest itself from the message by empowering the messenger. As Barack Obama famously exclaimed of his own philosophy, PD 2.0 was a strategy of 'leading from behind'.

Julian Assange, the founder of WikiLeaks and a former talk-show host for the Russian state-media giant *RT*, has been quick to point out the corporatist underpinnings of US influence operations. In his 2016 book *When WikiLeaks Met Google*, Assange quotes Google CEO Eric Schmidt as speculating in an interview with Assange that WikiLeaks' objective was 'to change the system in some fundamental way', but also to create a 'marketing prop. You want to have a marketing story.'[33] Seated alongside Schmidt and Jared Cohen (the creator of the Alliance for Youth Summit and co-founder of PD 2.0), his characterisation of the Americans and their approach to strategic communications is one of hopeless capitalism. Yet WikiLeaks, as an influence operation, pursued many of the same tactics.

By outsourcing information through third-party private news corporations,[34] Assange's team, like the State Department, sought to conceal through

Figure 1.4   Smart mob in Tunis. Nhar 3la 3mmar, Facebook, 31 January 2011

the veil of plausible deniability any direct involvement in the final product.[35] Moreover, both PD 2.0 and the WikiLeaks campaign to counter it, were predicated on the illusion of popular opinion. In the age of the Internet, it is no longer necessary to huddle together cooperators in secret as Professor Asch demonstrated in his Gestalt experiment from the 1950s. As PD 2.0 and WikiLeaks have shown, Facebook, Twitter and artificial commentators can achieve the same effect.

The Department of Deconstructionism

In the US, the idea that media – and social media in particular – may serve as a 'subjectless generator of structures', had been churning through the greater universe of the National Security Enterprise for over a decade.[36] Programmes like the Sociocultural Content in Language and the Metaphor project of the US Office of the Director of National Intelligence, or the Narrative Networks programme disseminated through the Defense Advanced Research Project Agency, were predicated on the notion that experts might 'systematically analyze narratives and their psychological and neurobiological impact' with little to no foreknowledge of the context in which they are created or the environment towards which they are projected. The task of defining 'substrates and mechanisms related to culturally relevant cognitions and behaviors' was believed achievable,[37] in part because the greater defence intelligence community had whole heartedly absorbed a theoretical perspective on the production of language and meaning that academics in the remote field of literary studies had been passionately debating for a century. 'The death of the author', as Roland Barthes famously decried, had created the opening for an analytical wing of that ancient enterprise in which studious readers with an eye for style could produce a text-centred approach to reading that promised to decode the intricacies of any written work with little to no regard for the context in which it was created or the life of the individual who created it.[38]

For the defence-intelligence industry the text-centred approach of narratology allowed for the prospect of an intriguing new product in the war of ideas, namely: an automated system of discursive analysis capable of sweeping the vast horizon of online chatter while instantly converting its finding into an actionable counter response in the form of audio-visual missives, leaflets, imagery, and talking points all uniquely designed to 'directly affect perceptions, emotions, behaviors, and tendencies for affiliation' within a target population.[39]

In the immediate years following 9/11, narratology surfaced primarily within certain rarified circles of the National Security Enterprise; beltway boardrooms where a combination of outside 'experts' and low-ranking officials began crafting the language for what would become a minor arms race among communications specialists, IT experts and the occasional political scientist jockeying for a piece of the pie in the increasingly lucrative war of ideas.[40] By the time of the Arab Spring the notion of narrative as 'an event without any subject'[41] had become engrained in the basic fabric of the greater US intelligence apparatus and nowhere more so than in the realm of counter-terrorism. In the fight against the Islamic State in Iraq and Syria (ISIS/ISIL), analysts routinely made reference to 'the narrative of ISIL' or even the 'the ISIL meme'. 'The ISIL meme builds upon the spread and acceptance in many extremist leaning Muslim circles of the narrative that Islamic lands, people, and the religion itself is under attack from Western powers', wrote one contributor for a large-scale, federally funded research study of the Islamic State's communication strategies.[42]

> Likewise, the narrative claims that despotic regimes in the Middle East are the fault of Western powers that thus should be attacked. Chechens and al Qaeda terrorists before the emergence of ISIL also argued that when their enemies used weapons of mass destruction, they too were justified in using such. And in the case of Palestinian and Chechen groups, women were encouraged to join the battle and an ideological basis was created based on fatwas that allowed the women to leave their families to join a terrorist group without asking permission of their male relatives. ISIL has coopted all of this into its meme.[43]

Assembled from an almost incompressible bricolage of keywords, flashpoints and CVE (countering violent extremism) jargon, scores of self-identified analysts like this one, often working exclusively in English, capitalised on what the authors of the *9/11 Commission Report* decried as the prevailing 'newsroom' culture among US intelligence contractors, who, by the 1990s, 'no longer felt they could afford such a patient, strategic approach to long-term accumulation of intellectual capital'.[44] The social-scientific or 'university culture' of old was part of a bygone era by the time of the Arab Spring.[45] In its stead appeared a kind of pseudo-analytical regime steeped in the language of narratology, however unconscious its authors appeared of such legacy.[46]

Independent of history and human action, it is the 'meme' or 'narrative' that spins its own meaning in such analyses; the author a Dionysian-like god 'making his promised presence all the more palpable to the sons and daughters of the West by means of his poignant absence'.[47] As Habermas famously observed of Derrida's poststructuralism, the underlying irony in the ostensible analysis of such 'archewriting' is that its predication on the occlusion of an author, or even 'human interest' in the general sense, prevails at the expense of precisely what it seeks to uncover.[48] 'The labor of deconstruction lets the refuse heap of interpretations, which it wants to clear away in order to get at the buried foundations, mount ever higher.'[49] In labelling or approximating the meaning of memes or narratives, the analyst is always at risk of precipitating their very purchase. Often times, as in the passage above, they do so through an obscurantist, sourceless discourse that portends in its own deconstruction, a fully-formed and ideologically vested perspective miraculously unbound from history.

The Department of Structuralism

The tendency among counter-narrative specialists to imagine narrative as being divested of a narrator is all the more ironic because it was precisely that same illusion that their counterparts in the world of public diplomacy sought to fabricate.

In his 2008 speech for the New America Foundation, James K. Glassman points to 'a short book' by Monroe E. Price, the head of the Stanhope Centre for Communications Policy Research in London, in which the author describes the work of the 'French deconstructivist philospher Jacques Derrida' and his 'tome' *Of Hospitality*. In fact a short collection of lectures from the late 90s, Derrida's book, as Glassman notes, has 'nothing at all to do with strategic communications'. But still, he asserts, the work holds fundamental lessons for the future of US influence operations. Derrida's book can help explain 'a major reason for animosity toward the United States: the view by others that we don't respect their opinions, that we do not actively listen and understand', Glassman says. His philosophy, 'in Price's reading', shifts the understanding of 'hosting' from one of control or ownership, to one of 'welcoming'. 'I like this paradigm: from the host as owner to the host as welcome', he asserts. It is 'a good description of Public Diplomacy 2.0'.[50]

By 'hosting' a network of communications operatives, or bloggers, many with little to no vested interest in US policy or any material connection to the United States beyond a few workshops or grant allocations, the US could wage the war of ideas in much the same way that the social media giants in

Silicon Valley had built their empires. Like Facebook or YouTube, Public Diplomacy 2.0 aimed to provide a 'platform for cooperation, mediation, and reception – a mode of being informed as well as informing'.[51] The strategy marked an evolution in the State Department's approach to public diplomacy. As Lina Khatib, William Dutton and Michael Thelwall observed, Glassman's predecessors oversaw what many perceived as an overly 'one-way' strategy in the government's approach to the war of ideas.[52] Contemporary iterations of Cold War vehicles like *Al-Hurra* TV or *Radio Sawa* functioned in the same space of 'white propaganda' as American infantrymen distributing leaflets in Afghanistan.[53] While more covert efforts eventually gained traction in Iraq and Afghanistan,[54] public diplomacy – including the work of the State Department's Digital Outreach Team – remained driven by a 'key strategic choice' to 'genuinely identify their posts'. [55] Across the greater intelligence community it could be said the US approach to strategic communications was defined largely by a reluctant set of 'societal attitudes' stretching back to the experience of World War I and revelations that the British had been quietly stocking apparent involvement in the war through a covert influence operation run through the Wellington House in London and Reuters news agency.[56] In the words of Colonel Dennis M. Murphy and Lieutenant Colonel James F. White of the Information in Warfare Group at the United States Army War College: 'countering American angst over the effective use of propaganda' had become as great a challenge for the US government in the war of ideas as was the rhetoric of the 'enemy' itself. To avoid the ire of politicians and the populace, Murphy's and White's recommendation in 2007 was the now familiar refrain: 'leading from the rear in the information war still gets the message told while avoiding any direct confrontation with democratic ideals'.[57]

Conclusion: Re-centring the Subject

In narrative it is 'verisimilitude and coherence' that remain, not truth. As Edward Said wrote quoting Vico: 'understanding or interpreting history is therefore possible only because "men made it", since we can only know what we have made'.[58]

The lines of association between PD 2.0 and April 6, or the bloggers who broke the story of the protests in Sidi Bouzid, were minimal if existent at all.

Yet the coexistence of their common perspective, their common objective and their common means presented a constellation of signifying elements sufficient enough for those hostile to the project of liberal democracy to assemble in their sights an 'antagonistic frontier' and to fashion, in turn, a counter-narrative.[59] Particles of truth created touchstones to what the leader of Egypt's 2013 coup and current President Abdel Fattah al-Sisi has described as America's use of 'Fourth Generation Warfare' (media propaganda, cyber-ops and soft-power diplomacy) to reshape the globe in its image.[60] But the bloggers' collectives, the corporate partnerships, the D.C. gatherings, may also be seen for what they were: independent alliances formed around a common cause, what moderate libertarians like James K. Glassman describe casually as 'freedom'.[61] For proponents of America's war of ideas this presents an easy exculpation. For those caught up by the narrative the Under Secretary sought to market and facilitate (see Chapters 5 and 6) – young writers clutching at an opportunity to travel and learn, activists looking for ideas – there was to be no excuse. Their fate was and remains tragically sealed.

If the failure of the world's respective politico-defence centres in stemming the tide of cultural conflict has taught us one thing it is that the Humanities ought to reclaim what was rightfully theirs to begin with, namely: intercultural understanding and the tools required for doing so, including narrative analysis and aesthetic critique. I have referred already to John Dewey's notion of 'art as experience'. And it is from this starting point, I believe, that critics of strategic communication may best imagine their necessarily two-fronted assault on the dual regimes of poststructuralism and defence-oriented data analytics. In both instances it falls to the humanist to re-centre the role of the subject and to re-focus collective interest on the human circumstances of cultural production.[62]

In reading the narrative dimensions of the Arab uprisings in Tunisia and Egypt over the first three years, this study seeks 'distinct historical practice, by real agents, in complex relations with other, both diverse and varying, agents and practices'.[63] The method of analysis falls generally within the rubric of cultural studies as Raymond Williams articulated it. The science it elicits, if any, is explicitly inexact. There is no tally of numbers or formula that indicate how one pattern of discourse or another precipitates response Y or supersedes response X. Rather, the methodology is one that refuses to gauge language in a vacuum – above all in the case of narrative, and especially in the context of revolution where the organisation of events into sequences, the assignment of antagonists, and the ascription of feelings are not merely exercises in communication but postulations on reality.

The FBI's exposure of the machinations of Russia's complex influence operation known as the Translator Project remains illustrative for this study. It is a reminder that even the most diffuse information, once crafted into narrative, necessarily contains perspective: a subjective vantage point on reality that is unique to the narrator's position in time and space. In the case of the Russian trolls working on behalf of the Trump campaign, that time and space was the Internet Research Agency in St Petersburg, 2014–16.[64]

Sketched in the unmistakable hue of 'black propaganda' wherein messaging is embedded covertly and, in the case of the 2016 US presidential election, via an elaborate network of trolls, fake news sites, false advertisements and memes, the Translator Project pursued a line of narrative that was identical to the one used during the Egyptian revolution and other points of political transition around the world since. Flooding Facebook with images of refugees, Hilary Clinton embracing women in hijab, or false advertisements for the Black Lives Matter movement, the Kremlin has sought to advance the narrative that immigrants, minorities and Muslims, in collaboration with naive globalists, the CIA and Georg Soros, are reshaping the ideological composition of the globe. In the face of this antagonistic frontier, Kremlin propaganda attempts to steer populations towards political campaigns that promise 'law and order' and a 'return to greatness' in the form of ethno-nationalism and economic protectionism. That the same trolls were used to disseminate the message in Egypt in 2011 and the United States in 2016 appears to be merely a point of indiscretion.

While often held apart from the political machinations of the West, the Arab world has witnessed a resurgence of identity politics that closely mirrors the wave of right-wing politics in Europe and elsewhere. The extent and nature of the trend is different across the region and elements of this dynamic have been in place for decades. But the counter-communications campaign directed against the initial democratic uprisings did much to reinforce them. In this way, it could be said, the information unleashed by WikiLeaks and promulgated through social media created a 'horizon of expectations' towards which pre-existent narrative identities could flow and ultimately find reinforcement.[65]

Scarred by instances of violence, terrorism and crime, Egypt and Tunisia – unlike Libya, Yemen, Iraq or Syria – evaded all out civil war. Yet, closer to the culture wars in Europe and the United States, the Arab uprisings set in motion for these countries an information war of historic proportions. The goal of de-escalating the ideological tensions inside these countries and between the Arab world and the West will entail a generational effort, but it is undoubtedly a goal worth pursuing as there is much real work to be done as well.

2

News of a Revolution

Introduction

On the morning of 25 January 2011, the neighbourhood of Sayida Zeinab in downtown Cairo was uncharacteristically quiet. By 8:00 am, café owners near the busy intersection of Midan Lazoghli and Khayarat Street were dragging chairs and tables inside. A police lieutenant waved passers-by from the street with his walkie-talkie. Leaning against the dusty entrance to the turn-of-a-century apartment block housing a rough-and-tumble coffee-house and low-cost print shop, a *bawwab*, 'Samer', gestured with his cigarette to the walled gates of the Ministry of Interior towering at the mouth of the street. 'People will come and shake their fists at the building', he predicted. Their chant would be simple – '*la*!' ('no!') – then they would leave.

It was a prescient observation. No demonstrators made it to the gates of the Ministry of Interior that day. Some 15,000 people reportedly turned out at Tahrir Square and elsewhere. But in a city of 20 million, such numbers hardly make a sound. At night the cafés reopened. The backgammon resumed. Hookahs burned slowly.

In the days and months to follow, this neighbourhood, just blocks from Tahrir Square, would be transformed by pitched battles, Molotov cocktails hurled from rooftops, tear-gas, water cannon, and an endless parade of disparate protests. But there existed no signs to that now famous destination of 11 February.

The revolution in the making became 'the eighteen days' and Tahrir 'Tahrir', as Margaret Litvin puts it.[1] But far from the discourse of 'waves' and 'counter-waves' and the sea of 'middle-range' analyses designed for broader debates within the realm of history,[2] the revolution, in Adhaf Soueif's words, was firstly a wide-eyed 'newborn' responding to the world around her and building consciousness through experience. A closer inspection of the weeks' daily discourse shows the degree to which the outcome of the revolution was anything but apparent to those watching it grow.[3] Certain trends in reporting among the local press defined the parameters of what would become the consensus history of the eighteen days. But these were negotiated patterns, formed and transformed by a range of conversant events and disruptions, many of which remain unexamined. Returning to the opening days of the Egyptian uprising, this chapter seeks to refocus on the constellation of signifying elements – the heroes and villains, plot twists and ironies – that helped define the emergent narratives of two principal entities in the greater body politic of the Egyptian intelligentsia – namely *Al-Ahram* and the liberal opposition newspapers *Al-Shorouk* and *Al-Masri al-Youm*.

Squaring the Circle

Emphasising print newspaper coverage of the Egyptian uprising may seem anomalous to the prevailing notion that the Arab uprisings of 2010 and 2011 were driven by online media. While not negating the evident importance of digital and satellite media, which I take up elsewhere, the focus of this chapter is premised on the notion that newspapers remain critical to our understanding of the uprising's early narrative dimensions. This is the case, I suggest, for two main reasons.

Firstly, from 27 January to 1 February – arguably the most crucial period of the entire eighteen days – the government succeeded in shutting down the Internet, turning off cell-phone coverage and darkening *Al-Jazeera*'s satellite feed. Looking back at my emails I can locate almost the exact moment the Internet began to sputter. My hurried last exchange with editors at the *Seattle Globalist* read: 'Inet totally jammed . . . Tmorrow's gonna be March 13 Big (in reference, erroneously, to the 15 February 2003 anti-war protest in New York and around the globe).

Over the next four days I dictated the stories I was writing over the landline in my apartment, a number I managed to pass along to my editors moments before the information blackout. Along with the rest of the country, local reporting was the only news I had access to.

But secondly, and regardless of these circumstances, it has always been

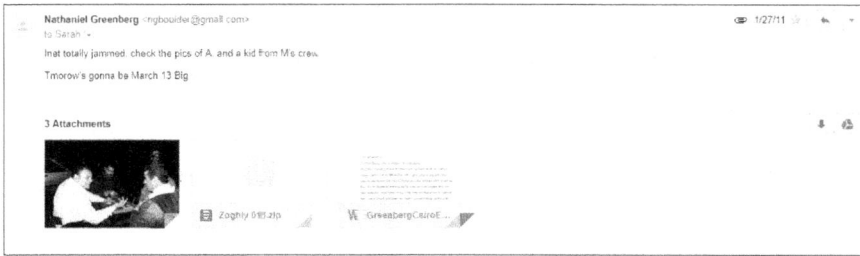

Figure 2.1 'Inet totally jammed.' Email from the author to editors, 27 January 2011

the case that it is the archive of the newspaper – not television, or radio, or online media – that most eloquently captures the imagined contours of a city's and a nation's daily life. As Benedict Anderson (quoting Hegel) observed, the 'extraordinary mass ceremony' of newspaper reading rests in the inherent temporality of its market appeal. To pick up a foreign paper with no prior background is a bewildering experience, something akin to entering a play already in its third act. But for a committed reader even the difference between a morning and evening edition reflects nuance, another peg in the ascending arc of some grand but unnamed drama.

'What more vivid figure for the secular, historically clocked, imagined community can be envisioned', Anderson wrote.[4] At a glance, the conservative organisation of life in Egypt would seem to challenge the relevance of such secular ceremony but still the premise holds true. No morning prayer (or evening broadcast) can be folded and pocketed and passed along verbatim to an acquaintance, a neighbour, a collaborator. In witnessing 'his own paper being consumed by his subway, barbershop or residential neighbors', the reader understands that the 'imagined world' he proscribes to is 'visibly rooted in everyday life'.[5]

Yet, simultaneously, this 'thingly element' of the newspaper reinforces and makes plain the aesthetic conceit of the newspaper's design.[6] 'The briefest comparison of headlines shows obvious variations in the story considered most important, in its presentation or angling, and in language and tone', wrote Raymond Williams.[7] Bending experience into headlines and headlines into leads, the aesthetic of the newspaper is at once concrete and open. Its meaning is bound to the materiality of the present, yet its reception depends on an unfixed horizon of receptivity; the millions of readers who may follow one thread or another, one section, one writer.

Magnified beyond the individual reader, the sphere of social communication in which a particular paper gains currency becomes in essence the congregation of Hegel's 'mass ceremony': a collective identity forged around a consensus faith in the objectivity, or narrative accuracy of the convening text. The dynamic is revolutionary insofar as it may or may not reflect the prerogatives of the state. As Habermas famously observed, the 'complex of the public sphere' has always been bound to the slackening and tightening of the freedom of press.[8] The dynamic is apparent in the case of Egypt where the proliferation and retraction of the free press has often reflected a cyclical-like formation and deformation of the public sphere. Yet, even amidst periods of expansion – in the late nineteenth century or the 1930s for example[9] – the quotidian newspaper served as a tool of the state, working to delineate the national imaginary while weakening if not forcing underground the role of print media in the discursive formation of the public sphere. It was not until 2004, with the licensing of private news media that the government finally allowed the creation of Egypt's first private daily, *Al-Masri al-Youm*. As Naila Hamdy observed, the reaction shared among officials was that the decision was 'one of Hosni Mubarak's biggest mistakes' as it became immediately apparent that the paper was functioning as a beacon for oppositional camps, in particular the Kefaya movement which had emerged the year prior and shared close ties with the paper's editorial leadership.[10]

Paralleling 'the al-Jazeera effect' on a national level,[11] the licit publication of independent papers – from the organ of the liberal opposition party *al-Ghad*, to the sensationalist weekly *al-Fajr* – created the prospect of true complexity in an otherwise static public sphere. But the dynamic was fragile. The founder of *al-Ghad*, Ayman Nour, was imprisoned months after the launch of the party's paper on charges of forgery (he also challenged Mubarak in the 2005 elections); the editors of *al-Fajr*, *Sawt al-Umma*, *Al-Karama* and *Al-Dostor* were prosecuted by a lawyer for Mubarak's National Democratic Party and sentenced to a one-year prison term and a fine of LE 20,000 for 'publishing false information likely to disturb public order'.[12] Scores of bloggers and individual journalists were persecuted or detained in the lead up to 2011. More radical voices, including the traditional organ of the Muslim Brotherhood, *al-Da'wa*, remained clandestine or published in exile.[13]

It was in this context – of fragile complexity – that the national imaginary of the public sphere was swept into the tempest of revolution.

A House Divided

On the morning of 25 January, the message emanating from the regime was one of assuredness. The front page of *Al-Ahram*, stacked atop the same dusty archives as always on the seller's mat, struck an unequivocal chord. Above a grinning portrait of Habib al-Adli, the Minister of Interior, the headline read:

> Terrorist organization including 19 suicide bombers attack houses of worship.
> The suspect Ahmed Latifi is a foreign officer.
> Security forces searched his house the day after the bombing.
> *The perpetrator of the crime is not Egyptian.*
> The suspect provides nothing about the incident.
> *No correlation exists between the Samlut and Alexandria attacks.*
> Church building authorities to reconsider regulations.
> *We will protect any gatherings exclusively committed to the expression of opinion.*[14]

Penned by none other than Osama Saraya, the newspaper's Editor-in-Chief who just months before had endured an international firestorm for defending the doctoring of a White House press release photo (*Al-Ahram* pasted an image of Mubarak erroneously walking in front of a group of international leaders attending talks with Obama), the communicative strategy of the state's flagship publication was on sharp display. The suspect and crime in question required no explanation. For a month leading up to 25 January, talk of terror dominated public discussion. The Two Saints Church attack in Alexandria on New Year's Day was the country's deadliest attack since the Luxor massacre in 1997. Officials and celebrities made great overtures to the Coptic community. Hani Shaker's '*Eid wahda*' ('All Hands Are One') extolling the unity of Copts and Muslims became one of the most popular songs on the radio. With little actual headway on the investigation, however, rumours about the bombing continued to abound.

The information disclosed by al-Adli's interview with *Al-Ahram* was not insignificant. The attack, he said, had been part of a transnational wave of violence targeting religious minorities across the region. Nineteen churches in all had been hit. The movement, which included the Army of Islam, the Gaza-based group associated with the suspect in question, was tied to the al-Qaeda affiliate the Islamic State of Iraq (ISI) and it included fighters from Libya and Tunisia, as well as militant Salafists in the Sinai. Many of the components that would coalesce into the transnational phenomenon of the Islamic State in Iraq and Syria (ISIS) were exposed by the Two Saints Church bombing.

Ironically, the ominous revelations were drowned out by the news that *Al-Ahram* was not reporting on that day.

Sitting alongside the seller's mat on the morning of 25 January was a wholly different view of reality. The independent newspapers *Al-Shorouk* and *Al-Masri al-Youm* were not only covering the first day of protest, but introducing to the Egyptian public a new framework of cognition.

> Day of Rage: Live rehearsal of demonstrations 'online' between government opposition and support . . . Police publish secret communications between activists and shut down headquarters of the Wafd Party . . . Social commit-tees incited by 'al-Sa'id' . . . April 6 distributes 15,000 new statements.[15]

With a clear backstory in place, the front page of the powerful independent newspaper *Al-Shorouk*, founded in 2009 by Ibrahim al-Mu'allim, a wealthy businessman with affinities to Mohamed ElBaradei's National Association for Change as well as the Muslim Brotherhood, set in motion a constellation of signifying radicals wholly distinct from the time-tested campaign of fear-mongering underway at *Al-Ahram*. The online 'rehearsal' referenced in the headline had occurred primarily on Facebook, which, in Egypt, with just 3.4 million users in a country of 80 million, was a rarified space still in Egyptian society. The major loci of this activity were even more obscure. 'Al-Said', a reference to the Facebook page 'We are all Khaled Said' had some 300,000 followers, a robust figure in the world of Egyptian social media but obscure in respect to the general population. The April 6 Youth Movement (*Harakat Shabab 6 Abril*), an online oppositional movement with ties to the Kefaya coalition and named for a 2008 general strike organised by the latter, had a following of some 80,000 members. It is uncertain how many actually resided inside Egypt when the uprising began.[16]

But the obscurity of the 'online rehearsal' – particularly for readers of the print paper – lent itself to the imaginative work of narrative. Tremendous mystery surrounded the origins of the uprising. None of the young men smoking hookah in Lazoghli on the night of the twenty-fifth had heard of the online groups. Indeed, none of the men I spoke with owned a computer and most with phones seldom went online. As the uprising progressed, government forces would exploit this uncertainty to great effect. The 'Internet youth', as a concept, functioned from the outset as an ambiguous signifier. Equally antagonistic as it was protagonistic, the label quickly became attached to a divergent range of ideological platforms.[17] Juxtaposing the exchange of words and pixels online with the movement of people in the street represented a major gamble on the part of the editors at *Al-Shorouk* insofar as virtual reality seldom comports with the concrete affect of experience. But the newspaper was not alone in crafting its opening salvo of reporting beyond the reality of the Egyptian street.

Al-Ahram too, on 26 January, disclosed a line of inquiry wholly removed from the actuality of the Egyptian scene. On the second day of protest, the state-sponsored giant published on its front page a vehicle lit ablaze with a large crowd amassing around it. The chaotic scene was not of Cairo, but Beirut, where, as the headline read: 'protests' (*ihtijajat*) and 'riots' (*idtirabat*) had widened.[18] This, combined with the other lead stories of the day – a recap of al-Adli's interview including its reception on Arab satellite channels and plans by the Minister of Foreign Trade and Industry, Rachid Mohamed Rachid, to reject a major trade agreement with the US – set the groundwork for what would become two indelible trajectories in *Al-Ahram*'s early discursive strategy: to magnify the menace of terrorism and instability while increasing the volume on calls for national solidarity. Offset to the bottom left of the front page the paper ran a story with the title: 'Thousands gather for peaceful demonstrations in Cairo and the provinces'.[19] Diminished by the raging scenes in Lebanon above and couched in a tone of banality, such language could not have been further from the discourse brewing in the media hubs of Doha, London or New York, or across town in al-Mohandessin for that matter, in the offices of *Al-Shorouk*. There, on 26 January, the headline was unequivocal: 'A volcano of rage sweeps through the streets of Cairo and explodes in Tahrir Square'.[20]

While *Al-Ahram* maintained an unassuming if not wholly dismissive tact in covering the first two days of protest, by 27 January *Al-Shorouk* was doubling-down on its reporting featuring a prominent bold and red-print headline that read: 'Random violence and acts of cruelty by security forces mark second day of rage'.[21] Amidst descriptions of rubber bullets and tear gas, the independent newspaper coined one of its lynchpin expressions for the first week of coverage: '*harb al-shawari*'. Cairo had spiralled into a 'war of the streets', its lead article read, pitting protestors against security forces and signalling a sharp divide in tactics between peaceful resistance and violent repression.[22]

For most of its readers this divide had been in place from the outset. January 25 was Police Day, the chosen symbol of revolt, and *Al-Shorouk*'s reporting was poised to exploit it. *Al-Ahram*, which, on 27 January, ran with

Figure 2.2   Riot police and paid government enforcers near Tahrir, Cairo, 28 January 2011. © Nathaniel Greenberg

the headline '4 Dead. 118 citizens and 162 officers injured' eventually turned to the same scenes of violence.[23] But where *Al-Shorouk* and soon *Al-Masri al-Youm* saw a 'war of the streets', *Al-Ahram* saw a security force 'committed to protecting freedom of expression within moderation' or 'without excess'.[24] No clear narrative was in place yet for the paper of record. Unlike the major opposition press, *Al-Ahram* had no backstory for framing the upheaval, nor had it articulated through its two-day-old coverage a clearly defined set of antagonists or protagonists in the way *Al-Shorouk* had done with the security forces and the enlightened trifecta of the Wafd, al-Said and 6 April. With the Friday of Rage approaching, however, the state media apparatus needed a different tact. Conveniently enough, a major new piece of information was about to surface.

Intifada Informatics

On the eve of 28 January, far beyond the shores of Egypt, external actors were hard at work casting an alternative narrative of the revolution. *The Daily Telegraph* of London was about to publish a story with an alarming new headline: 'Egypt protests: America's secret backing for rebel leaders behind uprising'.[25] The story, as mentioned in the previous chapter, appeared on 28 January. It detailed how the American Embassy in Cairo had facilitated covert training for a member of the April 6 Movement, how the individual met with House, Senate and think-tank staffers in Washington, and how, according to the American Ambassador Margaret Scobey, 'several opposition forces agreed to support an unwritten plan for a transition to a parliamentary democracy, involving a weakened presidency and an empowered prime minister and parliament, before the scheduled 2011 presidential elections'.[26] As late as 2016, Ambassador Scobey claimed to be unaware of this specific cable – indeed there were many.[27] But the response at the time was vociferous. After being picked up by the conservative American aggregator site the *Drudge Report*,[28] the article went viral generating nearly 1,400 comments in the first few days and being shared 23,000 times on Facebook. In addition to *The Telegraph*, articles from WikiLeaks appeared in the *Aftenposten* in Norway, *The New York Times* and on *CNN*. With the government having shut down the Internet, public response to the revelations inside Egypt were partially muted. Still the government moved swiftly to arrest the activist referred to in the article. At the time, the British paper redacted his name in an attempt to protect those around him. WikiLeaks offered no such courtesy when it published the memo on its website the following April.

The revelations unleashed by the shadowy online collective appeared calibrated to undercut whatever advantage the Internet had provided the so-called 'Internet youth', tying their funding, ambitions and organising directly to the United States. They did so not through coercion or intimidation, but, rather, as Sissela Bok once wrote, through the 'loss in judgement to secrecy'.[29]

The most powerful element of the revelations it could be said, existed not in the information itself, but in the headlines applied by WikiLeaks and the editors at *The Telegraph*. Such editing, along with the selection of samples and the timing of their release, was significant because the information contained in the memo, which pertained to 2008, was in essence archival and rather blatantly cherry-picked.

By utilising the news media, WikiLeaks was able to contemporise the story of the US's five-year Public Diplomacy 2.0 initiative. Collaboration also lent the information an air of objectivity. By flooding the comments section with trolls and spreading the information through viral online platforms like the *Drudge Report*, the operatives involved in the release were able to create the illusion of consensus. The effect (like that of the 'Asch Situation' described in Chapter 1) was one in which perceptive reality quickly gave ground to the coercive power of narrative.

The April 6 group, in reality, was a broad-based umbrella movement that included a multitude of political actors, many of whom had been working in relative obscurity for years to achieve modest democratic gains. Yet now the aura of secrecy had created within public discourse an imaginative horizon for the speculative constitution of new theories into the origins and objectives of not simply the April 6 Movement, but the entirety of the present unrest. The Egyptian government for its part was instantly absorbed by the 'euphoria of secrecy'[30] which, in turn, helped to align in its official discourse a diverse array of forces then colliding in Egypt. The 'online rehearsal' described by *Al-Shorouk* appeared far more complex in light of the WikiLeaks revelations. The DDoS attack on 26 January (claimed by the group Anonymous and an entity calling itself Telecomix which featured as its logo an upside-down Tunisian star and crescent plugged into a pyramid surrounded by lightning bolts) contributed to the impression that Egypt was under attack. Members of the American media (perhaps unwittingly) also contributed to the apparent deluge of external aggressors by pursuing eerily prescient lines of reporting. As early as 17 January, 'within hours' of Ben Ali's departure, Eric Trager, writing for the *Atlantic Monthly* reported on a minor gathering of 'approximately one hundred Egyptian activists demonstrating in front of the Tunisian Embassy in Cairo', before going on to describe their aspirations for a 'region-wide "Arab Spring", in which the "Tunisian scenario" would be replicated in Egypt and beyond'. Not disclosed in the article was the fact that at least one of Trager's interviewees in front of the embassy, Ahmed Salah, had participated in the Diplomacy 2.0 workshops named in the *Telegraph* leak. He also fails to disclose that his own Washington Institute for Near East Policy played host to some of the events.[31] The *Atlantic*'s unsettling façade of objectivity was further compromised on 27 January when the magazine circulated online the so-called Activists Action Plan simultaneously with its distribution in the streets.[32]

The Egyptian government's response to this apparent assault was swift. In addition to arresting the activist Ahmed Saleh, authorities disappeared Wael Ghonim and arrested scores of others associated with the 'online rehersal'.[33] (The co-founders of the April 6 group, Ahmed Maher and Ahmed Douma, were ultimately arrested too. The latter now faces life imprisonment.) ElBaradei, who, according to Ghonim, held a secret meeting with members of the Brotherhood and others from the Association for Change on the night of 27 January,[34] was placed on house arrest after marching towards Tahrir following Friday prayer at the Istiqama Mosque in Giza. The targeted abductions had minimal effect on the day's events, which quickly spiralled out of control. Yet, the following morning, with the towering offices of Mubarak's National Democratic Party still smoldering from an arson attack and the streets awash with remnants of the night's destruction, the cover of *Al-Ahram*, featuring the President's midnight speech, greeted readers with a resolute framework for understanding the apparent state of anarchy. The country was facing a 'plot' (*mukhatat*) to 'destabilize the establishment and undermine the law'. 'Reform is a retreat', Mubarak said:

We propose a bold plan for moving forward [look to the future] with new steps for relieving unemployment and improving standards of livings for

Figure 2.3   Mubarak's National Democratic Party Headquarters in flames, 29 January 2011. © Nathaniel Greenberg

citizens. We will not tolerate any act that compromises the safety of all Egyptians. Freedom and democracy will increase for the nation . . . as will respect for the independence of the judiciary and its rulings.[35]

On page eight, headlining its international section, *Al-Ahram* ran an article featuring the leak that appeared in the *Aftenposten*. The 'plot' to 'destabilize' the regime entailed 141 million dollars in support from Washington for 'pro-democracy' groups operating inside the country. The article singled out three cables, including a 2009 telegraphic memo sent by the American Ambassador in which she describes pressure put on her office by then Minister of International Cooperation Fayza Abul-Naga to cut funding for ten such organisations 'on the grounds that they were not registered NGOs'.[36] The images of mobs and chaos that now flooded its pages fell easily into the narrative of a country besieged. Hani Shukrallah, a founding Co-Editor in Chief of *Al-Shorouk*, recalled in a 2012 speech the 'mythology of foreign fingers' as it appeared on state television at this time.

> They filmed a young woman with her face blurred or darkened . . . like a prostitute. . . . And got her to confess on TV that she had been taken by the Americans, along with other members of 6 April movement . . . to a seminar on how to make a revolution . . . [it was] Jewish intelligence officers who gave the lecture,

said the woman in the recording which was looped over Nile TV.[37] Rumours about Kentucky Fried Chicken and Euros being distributed to protesters ran wild. Videos of foreigners 'infiltrating' protests went viral. As Naila Hamdy and Ehab H. Gomaa noted in their quantitative survey of the press from this time, some 420 out of a total 800 articles sampled from semi-official news outlets including *Al-Ahram* defined events through reference to a 'conspiracy'.[38] Regardless of the veracity of the narrative unleashed by WikiLeaks, however, the revelations, as it turned out, were a double-edged sword for the regime.

Protectionism, Populism and Military Deliverance

Along with the existence of a 'plot', Mubarak's speech sought to recuperate one of the most powerful elements of the youth coalition's narrative, namely, their claim to the proverbial 'future' of Egypt. The regime had a foothold in this rhetorical arena already. Cloaked in the language of SME ('Small and Medium Size Business') Gamal Mubarak had been sold to the public as the country's youthful and entrepreneurial future. *Al-Ahram* and other propaganda wings of the state had been pitching the political platform of the President's oldest son for years. And Mubarak's 28 January speech echoed many of those same themes. Here, however, the programme of 'increased employment and improved living standards' was welded to the promise of protection from an unnamed 'plot'. It represented an ideological doubling-down on the part of the regime, a wager made all the more evident by the ostensible smoke-screen of nationalism decorating the front page of *Al-Ahram*.

Offset to the bottom left of the 29 January cover, and adjacent to a photo of a protestor holding an Egyptian flag like a shield to the blast of a water cannon, appeared in bold print an announcement from the Ministry of Trade and Industry proclaiming a new era of 'consumer protectionism' and the promise to 'enforce rules and regulations on foreign trade and imports'.[39] In close step with the tone emanating from the President's Palace, the Stanford-educated minister, Rachid Mohamed Rachid, with the help of *Al-Ahram*, sought to package the narrative of securitisation with an economic message of protectionism, a position his office had been promulgating since the outset of the uprising. (*Al-Ahram* quoted the minister in its headlines on 26 January as saying, bluntly, 'we have no desire of entering into a trade agreement with America'.)[40] Yet, while attempting to self-characterise in the context of the demonstrations as anti-trade, anti-American and pro-Egyptian, it was the Ministry of Trade and Rachid himself that, until then, had most clearly defined the ideological posture of the Gamal Mubarak camp. Indeed, Gamal and Rachid had been virtually campaigning together since 2009 when they appeared on state television conducting a town-hall style meeting with 'small and medium size business leaders' and 'university students' to discuss the 'future' of the Egyptian economy.[41] Try as they might, the editors at *Al-Ahram* could not mitigate in clear terms the ideological overlap of American involvement in affairs of the state and the economic 'reform' packages associated with Gamal Mubarak's pro-business leanings.

Simultaneously, the arrival of an alternative protagonist in the military created for the oppositional press a new set of rhetorical flashpoints that would bend the narrative of the revolution away from its original trajectory as a youth-led struggle for democracy in the face of a corrupt and oppressive system towards a proto-nationalist story of military deliverance from a Western-backed regime with dynastic ambitions. For those in Egypt witnessing the shift after 28 January, there was no mistaking the sudden prevalence in the streets and marketplaces of the name Gamal Abdel Nasser. Sheparding this new narrative of the revolution into public discourse was one of the country's oldest denizens of public discourse, the one-time Editor-in-Chief of *Al-Ahram* and former confident of Nasser: Mohamed Hassanein Heikal.

Heikal's Coup

Beginning on 29 January, with a cover-page headline in *Al-Shorouk* ('Heikal speaks about what's happening in Egypt'), Heikal would appear regularly in virtually all of the country's major newspapers for the remainder of the eighteen days.[42] A household name in Egypt for decades, he emerged during the first week of protests as a historical guide, providing ready-made discourse for an otherwise eclectic chain of events.

At the core of his intervention was a single basic observation: 'the army has never been a tool for striking the people', he said in a 1 February interview with al *Al-Masri al-Youm*. 'Because the army is the army of the people. It is the foundation on which this country and its history were built.'[43]

Heikal's narrative caught fire. As checkpoints proliferated across the city, popular reception of the army was ubiquitous. Civilians and soldiers worked hand-in-hand to quash looting and suppress the air of chaos that gripped the

Figure 2.4   Arrival of the military in Cairo, 29 January 2011.
© Nathaniel Greenberg

Figure 2.5   Self-policing in the streets, 30 January 2011. © Nathaniel Greenberg

city. By the end of the week soldiers were screening happy demonstrators as they entered checkpoints outside the square.

For the original proponents of 25 January, as Amr El-Shobaki later observed, 'the power of [25] January expired the moment the popular uprising that had forced the state to reform, to do away with the project of succession . . . suddenly cried out "the military and the people are one hand"'.[44] For many others, it could be said, it was only after the rhetorical convergence of the military and the 'people' that the full revolutionary thrust of the uprising began to take shape.[45]

Attempting at first to align the regime with the narrative of military deliverance, Al-Ahram featured on its 31 January cover an image of Mubarak seated with Chief Intelligence Officer, Omar Suleiman, and the Army's Chief of Staff, Sami Anan. The headline read: 'President Mubarak meets with military leaders. Governorates directed to provide basic needs for citizens. Demonstrators in Tahrir demand reform and reject chaos.'[46] Juxtaposed to a photo of Mubarak seated shoulder-to-shoulder with the army chiefs was an image of orderly protesters marching in the Square. The montage exuded confidence, control and even – with Anan's outstretched arm pointing towards the Square – an air of coordination between the military and the marchers.

The illusion, of course, was unsustainable.

As demonstrators assembled for a 'Million Man March' on 1 February, the state flagship was again forced to change course. Revelations published by Al-Shorouk the day prior that the President's sons Alaa and Gamal had attempted to 'flee to London'[47] exposed the President to an aura of culpability. The news also compounded a central trope from the myth of succession that Gamal Mubarak and his cohort had ties to Western business interests.

While not mentioning the report directly, Al-Ahram's cover effectively sought to quarantine the information. 'A New Government without "Businessmen"' the headline read in direct reference to the President's innermost circle, including his sons.[48] The strategy of isolating the government – including the President – from the myth of succession and the trope of 'businessmen' reached a zenith of sorts on 3 February with the headline 'Millions come out in support of Mubarak'. Falling on the heels of violent clashes between protesters and pro-government supporters the day prior –

what became known as the 'Battle of the Camels' – *Al-Ahram*'s 3 February cover was deceptive. In strategic terms, however, the blaring headline was undeniably effective, serving to articulate in concrete fashion the coexistence of a parallel movement.[49]

The visual composition of the 3 February cover was similarly uncanny. Picking up a copy in the glare of early morning, the unwitting reader may have imagined he had stumbled upon a different '*Al-Ahram*', a copy of a copy perhaps from 25 January, the opening day of glorious protest. *Al-Ahram*'s 3 February cover – its first to show an aerial view of protesters en masse – was the expression of a different revolution, one whose seed had been planted not on 25 January, but, rather, 29 January, the day of the military's arrival. The exact components of this alternative revolution were far from fixed. (Mubarak would not replace 'Mubarak'.) But the framework of the new revolution was now in place.

Al-Gumhuriya, another pro-government daily, picked up on the strategy. Their 4 February headline read: 'The gains of the popular revolution', 'Detained youth will be released' and 'Gamal Mubarak will not run for office'.[50] Here was a convergence of the 'popular revolution' with 'the people'; a severing of Gamal ('the businessman') from the state and a simultaneous coupling of the myth of succession with the most potent antagonist of the eighteen days: foreign plotters. 'We respect our customs and origins', the headline concluded: 'do not believe statements by a foreign government which prove the existence of a foreign plot'.[51] Paraphrasing Suleiman, the warning (which topped the news at *Al-Ahram* as well) gained rhetorical currency within the context of the state's alternative revolution. Foreign actors and foreign tendencies (including Gamal Mubarak's 'business' leanings) had conspired to undermine the state. The people, of whom the military are a part, were demonstrating in defence of the nation, its customs and heritage.

This basic narrative structure would remain in place for the remainder of the eighteen days with headlines extolling 'dialogue' over 'revolt' and solidarity in the face of foreign interference.[52] By the time Mubarak finally stepped down (the reality of the Square and US pressure being, perhaps, too much to sustain), *Al-Ahram* was able to publish with minimal hypocrisy its now infamous 12 February cover featuring a well-trimmed everyman with an army-like brown-coloured shirt and arms outstretched cheering atop a throng of protestors. The headline read: 'The people topple the regime' ('al-sha'b asqata al-nizam'). It was the first time since 1981 Al-Ahram had gone to print absent an image of Hosni Mubarak.[53]

Chaos Control

Concurrent with the annexation of revolutionary sentiment by state-run media was the dissolution of the opposition's original constellation of antagonistic and protagonistic signifiers. Fuelled in part by the deceptive revelations of an April 6–Washington conspiracy, the shift in strategy on the part of Al-Masri al-Youm and Al-Shorouk interacted with and diverted from a complex range of material and rhetorical developments in and around the Square.

The obvious starting point for the narrative reversal described here was, again, 28 January, the Friday of Rage.

With police disbanded and the city on edge, citizens armed with sticks and machetes descended into the streets barricading neighbourhoods and breaking street lamps to detour outsiders. Reports of looting and armed assaults from across the city live-streamed on the state-run Nile TV and over the radio.

As I wrote at the time, this sudden retrenchment of social life into neighbourhoods, coupled with the absence of cell-phones and the Internet, intensified feelings that the city was under siege.[54] Local chatter fuelled by state-media reports warned of infiltrators. The evacuation of European and American embassies further exacerbated the President's warning of a Western plot.

Underpinning the story of the uprising was now the very real prospect of chaos, a sentiment vividly illustrated by the 29 January cover of Al-Masri al-Youm. Juxtaposed to the image of a police station in flames the paper went to press on Friday night with the full-sized image of a police officer, his face in anguish and shoulders lowered in a posture of defeat. The headline read:

> Curfew imposed on all governorates. The Army Called In. Street wars in Cairo. Police stations in the capital and governorates lit ablaze. Hundreds injured in fires and violent clashes between security and protesters in an outburst of rage.[55]

Figure 2.6   Neighbourhood guardians in Cairo, 29 January 2011.
© Nathaniel Greenberg

News of clashes between police and protestors, looting, and prison breaks filled Saturday's papers. By Sunday, 30 January, the aesthetic of revolution as it existed five days earlier in the pages of the opposition press now appeared altered beyond recognition. Gone were the characters of the April 6 Movement and Mohamed ElBaradei. Neither would reappear in the headlines for the remainder of the eighteen days.[56] In their place, rising from the masses, was the heroic image of a uniformed officer hoisted onto the shoulders of a jubilant crowd. The caption to *al-Masri al-Youm*'s cover shot read: '*al-sha'b wa al-jaysh: eid wahda*' (the army and the people are one hand).

With the original story of the uprising increasingly untenable the oppositional press moved to recast their dramatic ensemble. In the place of '*al-amn*' – a reference to the Interior Ministry's *al-amn al-markazi* (Central Security) – the dominant antagonist in the pages of *Al-Shorouk* from 30 January onwards was 'Mubarak' and by extension the more abstract signifier '*al-nizam*' (the regime). '*Al-amn*' appeared five times in the newspaper's headlines and subheadlines between 25 and 30 January and just once (a subheadline from 3 February) between 31 January and 12 February. '*Al-nizam*', in contrast, does not appear in the headlines of either the oppositional press or *Al-Ahram* until

30 January when *Al-Shorouk* features the term in reference to the now famous rhetorical axiom: '*al-sha'b yurid isqat al-nizam*' (the people want the collapse of the regime).[57] '*Al-nizam*' appeared in the headlines four times between 31 January and 12 February. *Al-Ahram*, in contrast, uses it just once, on 12 February, as part of a dramatic full-page cover: '*al-sha'b asqata al-nizam*' (the people topple the regime).[58] The rhetorical rise and fall of 'Mubarak' followed a pattern similar to that of the word 'regime'. The President's name is absent from the headlines of *Al-Shorouk* between 25 and 29 January. Beginning on 30 January, however, 'Mubarak' appears six times in the role of the antagonist.

As was the case with its shift in protagonists, *Al-Shorouk*'s rhetorical focus on 'the regime' (opposed to state security forces) reflected in part a pragmatic response to the reality of a country in chaos. Simultaneously, however, the shift in antagonists was also ideological, a move that sought to express in overt fashion solidarity with the arrival of the army. The meaning of the word '*nizam*' (regime), crafted in part by the state-run media itself, had assumed by early February symbiotic coherence with phrases like 'businessmen', 'foreign plot', 'London' and 'Gamal Mubarak'. With the critical exception of the President's son, these signifying radicals constituted, ironically, the very superstructure of the regime's own counter-narrative. Yet leveraged in opposition to the ascendant protagonists of '*al-sha'b*' (the people) and '*al-jaysh*' (the army), 'the regime' and finally Mubarak himself, as 'the head of the snake', gained clarity – a kind of unifying resonance bold enough to eclipse the cacophony of Tahrir.

In some respects the final emphasis on Mubarak and *al-nizam* (read also: 'the system' or even 'the order') tapped into the existing arc of a foundational narrative within the corpus of the resistance, namely, the 'myth of succession'.[59] In a more general way, however, there prevailed in the rhetorical shift of the opposition (as well as the discursive embrace of the uprising by the traditional organs of the state) a certain 'indeterminacy' of relation between the 'ontic content' of the opposing camps and the 'ontological' language used to describe their place within the unfolding experience.[60] To borrow from Ernesto Laclau's reading of populism, the language of Tahrir, by early February, embodied an outpouring of national sentiment so vast that the particularities of the original conflict became all but irrelevant. The 'nebulous no-man's-land' between 'left-wing and right-wing populism' had enveloped the press in a sea of grey. Absent from the headlines were the calls for housing subsidies, claims of police abuse, corruption, economic stagnation, 'bread, freedom, dignity'. In the place of these ontic particularities, was the greater imperative of narrative: an origin story (1952) and 'the sense of

Table 2.1   Rhetorical antagonists in the headlines of *Al-Ahram*:
25 January–12 February

Rhetorical **antagonists**	Number of mentions 25–9 January	Number of mentions 30 January–12 February
a [foreign] plot, Washington (Mukhatat)	1	1
Mubarak	0	3
Gamal Mubarak	0	2
members of the NDP	1	6
al-Mutazahirun/in (demonstrators)	1[a]	0
al-Muzahirat (protests)	2	2
fauda, nahb, takhrib (chaos, looting, vandalism)	0	5

[a] In some instances the role of a given signifier is not evident, e.g. 'clashes widen between protesters and security forces', *Al-Ahram*, 27 January.

Table 2.2   Rhetorical protagonists in the headlines of *Al-Ahram*:
25 January–12 February

Rhetorical **protagonists**	Number of mentions 25–9 January	Number of mentions 30 January–12 February
Ahmed Shafiq	0	4
Mubarak	3	1
al-amn (the security)	3	0
al-Hakuma (the government)	2	0
al-mutazahirun/in (demonstrators)	1[a]	5
al-sha'b, al-shabab (the people, the youth)	0	3
al-jaysh /al-quwwat al-musalaha (the army, armed forces)	0	3
Omar Suleiman	0	5

[a] On 26 January *Al-Ahram* published in a sub-headline: 'thousands turnout in Tahrir for peaceful protests', 'thousands' here being synonymous with 'demonstrators'.

Table 2.3 Rhetorical antagonists in the headlines of *Al-Shorouk*:
25 January – 12 February

Rhetorical **antagonists**	Number of mentions 25–9 January	Number of mentions 30 January– 12 February
al-shurta (the police)	1	0
al-baltagiya (thugs)	0	1
al-amn (the security)	5	1
al-nizam (the regime)	0	4
Mubarak	0	6
al-Hakuma (the government)	0	1

Table 2.4 Rhetorical protagonists in the headlines of *Al-Shorouk*:
25 January–12 February

Rhetorical **protagonists** in the headlines: *Al-Shorouk*	Number of mentions 25–9 January	Number of mentions 30 January– 12 February
Al-Wafd, April 6, al-Baradei	3	0
Heikal	1	4
al-sha'b (the people)	0	6
al-mutazahirun/in, al-mu'arida (demonstrators, opposition)	5	7
al-jaysh, al-quwwat al-musallaha (the army, armed forces)	1[a]	4
Omar Suleiman	0	6
Al-Ikhwan[b] (the Brotherhood)	0	1

[a] On 29 January, *Al-Shorouk* quotes the armed forces as saying 'we will not use force against the Egyptian people'. The phrase is indicative of the rhetorical crossover occurring between the liberal opposition and state-run media. While juxtaposed with *al-sha'b* in a dialectical manner, the content of the quotation 'we will *not* use force', positions the military as a collaborative force with the leading protagonist of the story: the people.

[b] The Brotherhood appears just once in the headlines of *al-Shorouk*, on 6 February: 'The Brotherhood agrees to dialogue before the President steps down provided it meets the demands of the masses.' The organisation featured more prominently in other publications including, notably, *Al-Gumhuriya*. Beginning on 6 February, their rhetorical function within the state-aligned publication was antagonistic (6 February) and protagonistic (7, 9 February). The organisation does not appear in the headlines of *Al-Ahram*.

an ending' (Mubarak's resignation).[61] The newspapers captured the aesthetic with broad strokes: 'The people move forward as Mubarak begins to retreat', read *Al-Shorouk*'s headline from 30 January.[62] The 'condensation' of the uprising's myriad goals to the sporting-like push and pull of the 'people' versus 'Mubarak' effectively abstracted the revolution beyond recognition, leaving in its place a clear dichotomy.

Paradoxically, as Laclau observed, the relentless imperative of an 'antagonistic frontier' dissolves the traditional fault lines between the left and the right, or, in the case of Egypt, the state and the opposition.[63] The effect became evident in the rise of Omar Suleiman. The country's chief intelligence officer emerged onto the discursive landscape with the headline (as reported by *Al-Shorouk*): 'Suleiman's proposals to contain the uprising of rage'.[64] An arbitrator, at first, his role within the narrative of both the state and opposition press became that of the hero; a steady hand working to placate and overcome a still unsettled horizon of antagonistic forces. As the security situation continued to unfold, his calls to 'dialogue' with the opposition crystalised what he was not, namely, Mubarak or, by extension, 'the regime'. The distinction, of course, was false. As millions of observers indeed recognised, Suleiman (while spending most of his career in the shadows) was an institutional stalwart. In the arena of populist rhetoric, however, his place within the unfolding narrative of the revolution stood in opposition to the generic antagonism of 'Mubarak'. By 11 February, the 'nebulous no-man's land' of discourse on the uprising had so thoroughly obscured traditional lines of division that the newspaper which first began depicting the events of 25 January as a struggle between security forces and pro-democracy leaders clamouring for justice, now featured in its headline the phrase: 'Mubarak refuses to step down and delegate powers to Suleiman. "Tahrir" rejects'. Juxtaposed to the entrenched 'stubbornness' of the dictator,[65] Suleiman, in some respects a more consequential figure of the 'regime' than Mubarak himself, was now in a position of defending 'Tahrir'. In the context of the narrative he belonged to he had become in effect the leader of the revolution.

Conclusion

No other medium depends on such 'profound fictiveness' of presentation as the newspaper, observed Benedict Anderson.[66] Indeed what other medium (apart from magazines perhaps) is so frequently collected to be framed and mounted like a work of art. With the 'illusion of the completed process' abandoned[67] – and power up for grabs – edited alternatives gain renewed authority. In such instances rhetoric precedes reality, which may or may not comport with the logic of the discourse in play.

Heikal's 'one hand, one people' narrative, conveyed through the steady tone of 'there and then', helped open the imaginative arena of Tahrir to the very antagonisms the revolution once sought to defeat. But the old lines of division were irrelevant by the time of Mubarak's resignation. What remained was the sheer 'anomie' of the country in revolt.[68] 'Hurub al-shawari'' (street wars), 'fauda' (chaos), 'al-baltagiya' (thugs). In the face of 'total disorder' 'whatever the Levianthan does is legitimate' wrote Laclau.[69] The rise to power of the Supreme Council of the Armed Forces (SCAF) appeared to express just that.

The former US Ambassador Margaret Scobey, among others, has said she was unsurprised by the outcome of the 25 January uprising. The Brotherhood and the military (like in 1952) had formed a tacit (if temporary) alliance. The youth activist movement, members of whom Scobey met while in Egypt, seemed underprepared for the task of revolution. Their platform was built on critique, not construction.[70]

Not accounted for in this estimation, however, was the scope of the counter-communications campaign levelled against the pro-democracy coalition and against US initiatives in cyber-diplomacy. The release of leaked documents from 2009, repurposed and reinforced by a wave of fabricated commentators (some of whom would surface again in the service of Russia's Translator Project), gravely undermined the credibility of the original 25 January leadership. Interspersed with the chaos of the first week, the disbanding of the police and the arrival of the military, the misinformation campaign surrounding ElBaradei, Ghonim and the April 6 Movement (each allied forces in the greater ideological sphere of US and Western liberal democracy) blended seamlessly into the consolidation of the antagonistic front surrounding the 'regime' of Mubarak's businessmen and Mubarak himself. In retrospect, this correlation of antagonisms seems unlikely. Within the broader field of public discourse they exist only in fragments. But measured anecdotally within the context of the opposition newspapers *Al-Shorouk* and

Al-Masri al-Youm, the convergence of antagonisms (chaos – the West – the regime) serves a definite aesthetic, one that celebrates order and security and the perceived national identity of the post-independence moment. Under al-Sisi, as I discuss in Chapter 5, this aesthetic would achieve full expression in the form of anti-Islamism, a narrative also fuelled by WikiLeaks and reinforced, somewhat paradoxically, by anti-US hysteria. The reordering of Egypt's strategic and economic alliances – from the US, Qatar and Europe to the UAE, KSA and Russia – followed suit. Whether or not the communicative and political disruptions were coincidental or coordinated, however, remains an open question.

3

Abu Ayadh: *L'Homme Revolté*

Introduction

Amidst the convulsions of a struggling economy and a parliamentary troika labouring to create a new constitution, there appeared, in the winter of 2011, a mercurial figure on the social landscape of post-revolutionary Tunisia.[1] Known by the name 'Abu Ayadh', Sayfallah Omer bin Hussayn al-Tunisi was a charismatic speaker with an extraordinary story. At lighting speed, he captured the spotlight of an increasingly crowded field of emergent voices. Although absent from the myriad talk shows or newscasts then flooding the airwaves and Internet, Abu Ayadh's presence became ubiquitous. From July of 2012 through to the end of the year seldom a day passed without mention of his group – Ansar al-Sharia – or al-Tunisi himself in at least one (but more often all) of the country's principal newspapers. Neither the most senior nor the most militant of the country's nascent, so-called Salafi-Jihadist current,[2] his persona, like his story, evinced a specific kind of ambiguity, a mystery of origin, of intention, even affiliation, that signified for moderate and conservative writers alike the terror of the Arab Spring.

Here was the story of the '*criminel utile*', in Michel Foucault's terms, an almost mythological creature emergent from the dungeons of the state whose marginality to society would serve in expanding the state's arc of surveillance.[3] Unaware or unconcerned by this paradox, for the thousands he rallied – including many who heeded his call to jihad in Libya, Syria and Iraq[4] – Abu Ayadh and the spaces he occupied (prisons and mosques, impoverished banlieues of the capital, Gabès, Kairouan, Sidi Bouzid) were scenes of resistance; an open spectacle of the revolution itself. Mobilising a vast network of allied voices, Ansar al-Sharia's umbrella organisation, al-Qaeda, directed towards the Tunisian public sphere a full-scale communications operation in the form of *da 'wa*, their ostensible objective being to refute 'false narratives' and to position Ansar al-Sharia and its leader at the centre of an imaginary and ascendant vanguard. For those in the mainstream media he was a counter-revolutionary figure. His rhetoric contained within it the very antithesis of what made his critique possible: a language of order and restraint and a renewed call for restrictions on civil liberties in the form of sharia. Framing his place in society while mitigating his impact on the political scene became a major challenge for the Islamist party in power, Ennahdha. Meanwhile information operations beyond the shores of Tunisia looked to exploit the cultural fault lines delineated by his movement; a phenomenon made all the more explicit by the stunning if ideologically complex missive by WikiLeaks that Abu Ayadh was also an informant in the US-led War on Terror.

Out of the Past

It is important to note, firstly, there exists far more hearsay than fact surrounding the description of Abu Ayadh. Apart from a few disparate appearances before a camera only a small handful of interviews materialised publicly, one of which was recorded by a reporter for the country's largest radio station, *Mosaique FM*, but subsequently banned from airing. Most of the available biographical information on him was gleaned from a single interview posted on the website *Muslm.org* in early 2012. Additional anecdotes from a disassembled Facebook page placed him in the heart of the notorious 'Londinstan' crowd of the early 1990s, and in the immediate circle of Shaykh Abu Qatada, one of the principal recruiters for al-Qaeda.[5] Journalists trying to summarise his story typically began here before recounting his *périple* through the myriad hot spots of the Islamic world, 'Syria, Afghanistan, Iraq'.[6] His arrest in Turkey, in March 2003, and his subsequent transfer to Tunisia where a military tribunal-like court sentenced him to serve forty-three years, surfaced as the factoids most often associated with his name. Like many incarcerated under Ben Ali's notoriously strict anti-terrorism laws of the early 1990s, the exact charges against him remained undisclosed. But in the 2012 interview, Abu Ayadh said he had arrived in Turkey after fleeing fighting in Jalalabad, Afghanistan, where he and other Arab fighters who vowed to defend Osama Bin Laden following the dispersal of the Taliban faced the army of the

Northern Alliance in November 2001.[7] He had already indicated this history in a previous speech where he noted, in less than subtle fashion, that his location in Jalalabad placed him in direct proximity to the core leadership of al-Qaeda. For critics and followers alike, this anecdote became something of an open secret in the narration of his story. But the public history of his involvement with al-Qaeda, in fact, grossly underrepresented his actual level of involvement with the Jalalabad circle. As disclosed by WikiLeaks' so-called 'Guantánamo files', Abu Ayadh had been charged by al-Qaeda leadership with establishing a guesthouse for Tunisian recruits, the largest faction of foreign Arab fighters in Jalalabad. From there he was thought to have created, in early 2000, the Tunisian Combat Group (TCG), a transnational organisation that included running a major drug ring between eastern Afghanistan and Italy (where members of the group were wanted by the Italian police). On 9 September 2001, al-Qaeda military commander Abu Hafiz al-Masri reportedly directed members of the TCG to assassinate the Northern Alliance commander Ahmad Shah Masud, an operation designed to signal commencement of the 11 September attack on the United States.

This history far overshadows the known story of his life in Tunisia. As detailed in a report by the international human rights organisation Al-Karama in 2007, Ben Hussayn, born in 1965, originally fled Tunisia after a wave of arrests targeting militant student activists in 1987. He travelled to Morocco where he married before moving to the United Kingdom in a bid to seek asylum. At the time of its filing, Al-Karama claimed Ben Hussayn, then detained at the Mornaguia Prison in the Governorate of Manouba, Tunisia, was suffering from asthma and chronic kidney disease and had not received treatment 'for months'. The report indicated that his mother feared for his life. Her only visit in 2007 was abruptly cut short. 'His last cry', his mother said, 'was "somebody help me".'[8]

As described in a United Nations Security Council Sanctions List from 2018, Abu Ayadh was released from prison in 2011 as part of the general amnesty adopted by the transitional government just six days after the departure of Ben Ali, on 20 January 2011.[9] Having served a total of eight years, he returned to the streets of Tunis – the banlieue towns of Ould Ellil, Douar Hicher and Sidi Thebet – where he began promulgating an energetic *da'wa* or missionary campaign.[10] He sought financial support abroad, including in Oman, where he travelled in person to solicit funds. This trip, described by a reporter for *al-Sabah*, failed. Others, evidently, did not.[11] He organised elaborately staged charity events and supply caravans to impoverished areas of the country. And he delivered speeches, most of which were uploaded to YouTube, often with flashy introductory montages and audio overlays. The influence of Ansar al-Sharia was compounded by an ostensible wave of Salafist activism, with loosely affiliated sheikhs and other community organisations gaining control of some 400 mosques across the country.[12] Capitalising on the collapse of the country's strict Internet censorship laws, Ansar al-Sharia created a solid online presence with neatly choreographed demonstrations and public exhibitions carefully calibrated for online distribution. Assistants and attendees often wore bright orange vests symbolic of Iraqi war and Guantánamo prisoners. Stages and venues were surrounded broadly by the ubiquitous black and white flag of the *shahada*. Ansar al-Sharia events even included a sporting exposition, as witnessed during a rally in early September 2012 that featured a performance of classical stick fighting. The sportsmanship of the event was far overshadowed by the aesthetics of the spectacle. As evidenced by the pace of the well-advertised performance and the sea of cameras surrounding it, the 2012 exhibition was designed to maximise the symbolic imagery of the group – including visual references to the Palestinian struggle with the black and white checkered keffiya, and invocation of the club of the *futuwwa* – an early Islamic reference with roots in the story of the Prophet Ibrahim smashing the idols.

In exultation of such unprecedented freedom, the crowds at these demonstrations indulged in impassioned new chants that epitomised the postmodern pastiche of Salafist rhetoric: '*khaybr khaybr ya yahud* . . . the armies of Muhammad are coming', or '*Obama, Obama, koluna Osama!*'[13] While some were planned months in advance, others emerged in reaction to happenings in the public sphere: the screening of a controversial film, university protests, or fishermen selling black-marketed liquor at a portside market. At the group's largest rally, in May of 2012, in the ancient city of Kairouan, Abu Ayadh distributed tickets emblazoned with the group's insignia and encouraged his audience, with no shortage of irony, to behave as 'political police' (a term used by Ben Ali's anti-terrorism forces) and to enforce sharia wherever they saw fit. A string of violent episodes ensued over the summer

Figure 3.1   From the former Ansar al-Sharia YouTube channel, 2012

including raids on the capital's upscale neighbourhoods and the ransacking of an art exhibition perceived to be displaying blasphemous imagery.[14] The following September, in the wake of two controversial videos, both virulently denounced on the group's Facebook page, American embassies throughout the region were besieged by protests. Both of the attacks in Tunisia and Libya (where the American Ambassador was killed) were attributed to Ansar al-Sharia. Abu Ayadh, as the emir of the organisation's largest branch, quickly became one of the most wanted men in the world.

The Phantom Space of Terror

This is only a general summary of the disparate events that began to accumulate around his name. Already a figure of intense speculation, his story became yet murkier following the dramatic events of 11 September 2012. In the days following the attacks on the embassies (see Chapter 4), Abu Ayadh fled to the downtown Tunis mosque Al-Fath. Though international pressure mounted on authorities to arrest him, he was reportedly allowed to escape after police withdrew their positions citing traffic congestion. Two years later, the former Minister of Interior, Ali Laaraydh, explained that he had allowed Ayadh safe passage from the mosque for fear of violence between police and the hundreds of demonstrators that had gathered outside to support him.[15]

Speculation intensified following his disappearance. As though tracking his imaginary course out of the country, Tunisian media began running a series of stories on smuggled arms, even chemical explosives, along the Algerian border where it was believed he had fled.[16] Other reports surfaced that Ayadh had been allowed to relocate to a house in central Tunis, placing him in direct proximity to the Ministry of the Interior.[17] In his absence, new leaders emerged and rallies continued, though in much reduced fashion.[18] The assassination of the labour leaders Chokri Belaid and Mohamed al-Brahmi, in the spring and summer of 2013, drove remaining members of Ansar al-Sharia underground. More hardened militants reportedly fled to the Chaambi Mountains along the Algerian border where they were said to have joined forces with the Brigade of Uqbat Ibn Nafiha, a group Laareydh claimed had aligned itself with al-Qaeda in the Islamic Maghreb (AQIM).[19] All the while speculation about Abu Ayadh's whereabouts intensified. The online news site *Kapitalis* and the Algerian newspaper *al-Khabar* placed him in the Libyan capital Tripoli where he was said to have evaded airstrikes targeting him and his partner in hiding, the yet more infamous Mokhtar Belmokhtar of AQIM. Belmokhtar's death had been previously announced by Algerian authorities on at least three occasions. As late as October 2014, Abu Ayadh was reported to be smuggling illicit money wrapped in chocolate bars and date boxes to his brother in Tunis using a young girl as courier.[20] In November, Tunisia's Ministry of the Interior told reporters that he had surfaced in Derna, Libya.[21] Two months later, on 15 January 2014 (the third anniversary of the first day of protest in the eastern Libyan city of al-Bayda), Ansar al-Sharia's media network in Tunisia released a recorded message of Abu Ayadh pledging his allegiance to the 'Mujahidin brethren in Syria'. The announcement came just days after the deputy Libyan Minister of Industry, Hassan al-Darouei, was assassinated outside the city of Serte, soon to be the capital of the Islamic State's stronghold in Libya.[22] Abu Ayadh's announcement of a trans-Mediterranean alliance with fighters in Syria preceded Abu Bakar al-Baghdadi's formal declaration of a global Caliphate from the pulpit of al-Nuri Mosque in Mosul by five months.

Foreign Phenomena

Despite his physical disappearance, Abu Ayadh, as a historical reference, remained entrenched in Tunisia's public discourse. In mid-October 2014, the High Independent Authority for Audiovisual Communication (HAICA) imposed a month suspension on a television programme that hosted a guest reporter who boasted casually about his friendship with Abu Ayadh.[23] In an extended Op-Ed on the episode in the Francophone daily *Le Temps*, writer Faouzi Ksibi defended the decision, on a 'semiotic plane', suggesting, in essence, the broadcasting station itself, and not simply the reporter, was complicit in a communicative battle waged to gain empathy for 'the terrorist'. This narrative – of guilt by complicity towards Abu Ayadh – culminated in January 2015 with Laaraydh appearing before a federal judge in Tunis to testify about his decision to not arrest Ayadh while he was holed up in the Mosque al-Fath.[24] It was a charge, in essence, that had been levelled against the Ennahdha establishment since their taking power in October 2011. By 2015, the court of public opinion had converged with the judiciary. The charge of association, let alone complicity with Ayadh, implied little less than an act of treason.

In *Le Temps*, as with its Arabic counterpart *al-Sabah*, discourse surrounding Ayadh largely followed two major trajectories. The first concentrated on the coercive danger of his movement, that members of the highest ranks of government were susceptible to his ideology, if not complicit in its objective. The second focused on the kinetic, almost phantasmal-like quality of his character. Since his disappearance in 2012, seldom an article appeared without recalling his figure 'in flight', or 'fleeing', a description, which, as with his escape from the Mosque of al-Fath assumed almost mythic overtones. The vicinities of his textual existence were almost exclusively 'mountains' and 'borders', 'Algeria', or anti-places: 'vast movement between cities', 'caravans', 'online'. Such imagery both enhanced the invasive character of his narrative as something vaguely foreign and the destitute nature of his being, always scrambling, on the run, 'circulating incognito' across the Sahel.[25] These liminal allusions intersected with the earliest descriptions of his reemergence onto the streets, from prison to the depths of the city's periphery, and to his ultimate disappearance, 'smuggled' (*tahrib*) from the Mosque beneath the watchful eyes of the public and by the complicit hands of a corrupt government official.

Rached Ghannouchi, the leader of the country's largest Islamist party Ennahdha, who faced a raft of scrutiny in 2012 for ostensibly condoning the

Salafi-jihadist actors as 'sons' of a common cause,[26] sought distance between his movement and that of Abu Ayadh by emphasising the extra-national dimensions of the organisation. 'They're waging a war against the Tunisian revolution and against the Arab Spring', he said in a 2015 interview with Olivier Ravanello. 'I have no doubt that they are receiving external support. Terrorism is *"transfrontalier"* and *"transtemporel"*,' he added. 'It cannot limit itself, *intra muros*, because the rewards are often elsewhere.'[27]

Much of the discourse on Ayadh, of course, stemmed from his own rhetoric. In the most substantial such example, the banned interview with *Mosaique FM* reporter Nasser al-Din Ben Hadid, Ayadh described the jihadists' effort in Mali as part of a vast conspiracy born of myriad nationalities: 'Nigerian, Somalian, Chadian, Nigerien, Senegalese, Moroccan, Mauritanian, Tunisian, Libyan, and Egyptian'. He invoked as the stage for his struggle 'the Maghreb'. And of the hostage standoff at the Algerian gas complex InAmenas, a broad strategy of terror equally concatenate as the Arab Spring itself.[28]

This transnational emphasis, amplified by the post-revolutionary emergence of myriad eponymous cells from Morocco, to Libya, Tunisia and Yemen, aligned with the regional schemata of al-Qaeda in the Islamic Maghreb that openly supported the *da'wa* movement as a method for effecting public opinion. Yet the discourse on Ansar al-Sharia contrasted sharply at times with the group's otherwise mundane public messaging campaign which emphasised, as their name suggests, the need for conservative social reforms. A lengthy 2012 post from the group's now dismantled Facebook page conveyed their ideology explicitly: 'following the flight of Ben Ali some extremist [sic] currents (*tayyarat*), as well as atheists and secularists have launched repeated attacks against the sanctities (*muqaddasat*) of Muslims in Tunisia', the message read.

> These media attacks have infringed [on the rights] of the Companions [of
> Mohammed] the Messenger of God: peace be upon him and God bless
> them. [The insults are as follows]:
> The mocking of the *Niqab* and the *Hijab* and calls to prevent the *Niqab* in
> educational institutions.
> The mocking of prayer. The legal [sic] offense of our Prophet: PBUH.
> The advocating of normalization with the Zionist enemy.

The advocating of pornography and the decriminalization of adultery.

The advocating of homosexuality (sodomy and lesbianism which are
    falsely labeled as 'homosexuality') [sic] and the decriminalizing of the
    act.

The legal [sic] offense and disparagement of the Book of God.

Offending the Sacred by showing the infamous video tape [*Persepolis*].

Last but not least, infringing on the safety of the mother of Aisha's
    faithful: may God bless them.[29]

Preceding the mass protest at the University of Manouba in June 2012, where Salafi activists shuttered the campus and attempted to raise the black flag of the *shahada*, the group's rhetorical emphasis on the minutiae of local 'offenses' signalled what Western analysts like Aaron Zeilin described at the time as a definitive shift from the 'unipolar global jihad of the past decade' to a renewed 'multipolar jihadosphere, similar to the 1990s'. The key difference was that in the past 'jihadis thought locally and acted locally, while many now talk globally and act locally'.[30] The rhetorical motif 'Tunisia of Kairouan', in reference to the country's ancient city of Islamic learning, reflected this dynamic succinctly. In a subsequent Facebook post from the same period, Ansar al-Sharia assailed the troika government of Tunisia for 'keeping their pledge of being an alternative base for the Western crusaders and the State of Israel'.

They pass off other slogans, scheming for the gains made after the flight of
Ben Ali and for the exclusion of Islamist movements from all arenas of life.
For all of this we ask our people to take heed of the plotting against them
and against their sacred values (*muqaddasat*), to seek one another's hand
and to stand shoulder to shoulder in defense of our sacred Islamic values.

Not unlike claims by far-right religious nationals in the United States, Europe or elsewhere, Ansar al-Sharia's protest in early 2012 centred on the jurisdiction of secular authorities in lieu of a perceived divine ordinance. While ostensibly general in scope the position emerged in response to the national debate surrounding the work of the Constituent Assembly which was in charge at the time of drafting a new constitution. This work captured the attention of senior leadership in the world of the North African 'jihadosphere'. Much of

Ansar al-Sharia's activity in 2012, for example, appeared to stem from a well-circulated fatwa posted by Shaykh Abu Muslim al-Jazairi, an AQIM-linked ideologue and Sharia Committee member of the online forum *minbar tawhid wa jihad* (founded by imprisoned Salafist cleric Muhammad al-Maqdisi) who became a prolific voice in denouncing the creation of a new constitution. In response to a question posted to the forum, al-Jazairi wrote,

> As you have mentioned, the secularists strive to distort the identity of beloved Tunisia. This initiative is complementary to the procedures initiated by Bourguiba and developed by Ben-Ali. Thus, watch out for these malicious acts which are not new, because they want to stabilize this old malice [sic] with the cloak of reforms, and the maintenance of acquirements [sic].[31]

The statement, which evolved in response to concern among conservative activists that the Constituent Assembly would attempt to alter the religious identity clause of Article 1,[32] attempted to bridge the natural divide between practical questions concerning a media campaign and proposed petition to support the Article (using 'electronic websites, and Facebook') and the supranational imperatives of the greater al-Qaeda agenda. The first initiative, according to al-Jazairi, was permissible insofar as 'the disclosure of the symbols of secularists' was 'a collective duty and the brothers are required to do it to support the righteous and stop the vice [sic]'. The second, however, concerning the collection of signatures, fell outside the purview of tolerable action as it 'fit the methods of democracy in achieving the principle of realizing the majority opinion'. Any effort at building consensus, even if opposed to a perceived secular threat, would affirm the legitimacy of a political domain outside the realm of God.[33] As Abu Ayadh exclaimed in his speech at Kairouan: the mujahidin fought for '*hukumat Allah*' while the secularist desired '*hukumat al-jahiliyya*'.[34]

Following the massive turnout at the Kairouan rally on 30 May 2012, Ansar al-Sharia continued to stage public protests around the country. On 10 June, a demonstration in front of an art gallery in La Marsa became a scene of violence when activists ransacked exhibits thought to be displaying blasphemous imagery. Police posts and other galleries in the upscale neighbourhoods of Carthage, Kram and Essidjoumi were targeted at the same time.

Liquor stores were raided in places like Sidi Bouzid. In Jendouba a security outpost and local business were firebombed with Molotov cocktails. All of this, however, was largely a prelude to the outburst of violence following news of the online film *Innocence of the Muslims*. Those demonstrations led to the eventual attack on the American Embassy in Tunis as well as an adjacent school on 11 September 2012.

War of Narratives

While the leadership of Ansar al-Sharia typically claimed their activity focused on promoting a 'peaceful agenda', as several outside scholars have suggested,[35] media coverage overwhelmingly emphasised the frequent incidents of violence associated with their name. The trend became a substantial point of irritation in much of the group's own public discourse where the denunciation of 'falsehoods' was perceived more as a spiritual struggle in line with the work of *jihad* than a political effort to repair public opinion. The focus was a major point of concern across the jihadosphere generally as activists like those associated with Ansar al-Sharia pursued *da'wa* and sought greater visibility in the space provided by the absence of the authoritarian regimes. Quoting a verse from Surah *al-An'am* ('And thus do we detail the verses, and [thus] the way of the criminals will become evident'), al-Jazairi noted in his fatwa that the denunciation of perceived falsehoods was a divine order. The focus became equally pronounced in other communiques from AQIM at this time. Salah Abu Muhammad, the group's media official, released a lengthy letter in 2011 denouncing a single article from the London-based newspaper *al-Hayat*, where, as he claims, the author had fabricated a series of claims attributed to the 'leader' of AQIM.[36] The communicative front emerged not simply as a mechanism for self-defence or justification, but, as Abu Muhammad claimed, the seed of revolution itself: 'You cannot reduce the reasons for the revolutions down to one factor', he wrote. Still, it was evident that 'some of them are related to *da'wa*, media, politics, jihadist, economic, and social reasons'. As a careful system of deliberation was typically exerted on such communications, there is little doubt his ordering of causes was not without reason.[37]

Active jihadists in the post-revolutionary context arguably prized accurate communications above virtually all else. Such was the tone of the captured document sent to Salah Abu Muhammad by al-Qaeda Central ideologue Atiyatallah Abu Abd al-Rahman in 2010 where he notes: 'in sermons and statements applied to *the real world*, we must ensure – in order to attract the community to stand with us against the worldwide enemy – that we repeatedly state that our greatest enemy is the Americans'.[38] Echoes of the organisation's communicative foundation in oral tradition – a practice inherent to both its cultural identity and its vulnerability to law enforcement – become abundantly clear in its discourse on veracity. Driven largely by a relentless focus on 'malicious deception', a strategy, as one AQIM ideologue extolled, that was 'meant to divert the hearts and minds of the supporters of truth from following the battle between the believers and nonbelievers',[39] AQIM, Ansar al-Sharia and virtually all organisations fashioned in the pursuit of religious zealotry exhibit what Sissela Bok once described as the 'immobilizing impatience with all that falls short of the "whole truth"'.[40] Truth – '*altheia*' – in the pre-Socratic tradition, she notes, was understood as 'encompassing all that we remember', of what we salvage from the 'river of forgetfulness'. It is, in this sense, all that exists.

> The oral tradition required that information be memorized and repeated, often in song, so as not to be forgotten. Everything thus memorized – stories about the creation of the world, genealogies of gods and heroes, advice about health – all partook of truth, even if in another sense completely fabricated or erroneous.[41]

Western commentary on al-Qaeda linked movements has typically sought to cordon off jihadist rhetoric as an exemplar of madness. As Philippe-Joseph Salazar so lucidly observed, 'on the mental and physical maps of our vision of the world', jihadism occupies a kind of no man's land beyond the pale of reason. 'We can consult the map', he writes sardonically: 'There exists a demented territory, around Mosul.'[42] Particularly in the context of the information age organisations like al-Qaeda fit the description of a kind of 'schizophrenia'. Their imagined 'whole truth' and the ever-present cacophy of mainstream half-truths seem pathologically incompatible. Not accounted for in this estimation, however, is the fundamental belief structure underlying the movement's impulse for narrative coherency.

Decrying media reports as an act of 'Satan's handiwork' is an act of faith within the Salafist tendency and they are not alone among world religions.[43] 'Jesus calls Satan "the father of the lie" (John 8.44)' observes Bok in quoting the German ethicist Dietrich Bonhoeffer. 'The lie is primarily the denial of God as He has evidenced Himself to the world.'[44] But the technological and intellectual affront to absolutism in the digital age also elicits the need for reciprocity – for making an object of belief evident, for 'bringing it to life'. As Bok observes, this impulse is inherent to the creation of art. Works of terror, similarly, emerge as an attempt to materialise the otherwise immaterial nature of 'altheia'. Aesthetics, like narrative, provide a scaffold upon which the imagined world may enter the material one and communicate, therefore, that which has been suffocated by the silence of non-belief.

Ansar al-Sharia became a vital organ for Sunni jihad in this way as their public exhibitions and bold usage of social media – including a dedicated Facebook page and YouTube channel – broke new ground in confronting the 'pickaxes of the non-believers' with the 'fortress of truth'.[45] Yet, such 'truth' or *altheia*, in the pre-Socratic sense, thrived only by way of its immateriality, as myth and story, illusion and emotion. It is not surprising therefore that the public manifestation of narrative, of the performance of story in the form of diatribe, recitation and poetic allusion became the group's predominate model of *da'wa* as well as its most powerful vehicle for recruitment. In one of the most important studies of the group conducted during the brief window of access provided by the political opening of 2011, Fabio Merone and Francisco Cavorta found among young initiates in Tunisia that 'the stories of the al-Qaeda *mujahidin* reminded of the epic of the *sahaba* (the first group of Muslims)'.[46] The lore of Abu Ayadh's proximity to the mujahidin of Afghanistan, huddled with the 'Old Man on the Mountain' beneath the cloak of conspiracy invoked in-and-of-itself the material realisation of an otherwise distant fantasy.[47] Fittingly, Ansar al-Sharia's visual aesthetics were radically simplified, pared down and, at once, abstract. They zealously adhered to prohibitions of the figurative, to the '*mysterium tremendum*' of the sacred and the absolute.[48] The group's signature orange prison vests, which reappear in ISIS execution videos staged against the windswept desert, the theatre of Palmyra, the shore of Cyrenaica, while responding to Abd al-Rahman's message that *da'wa* ought, for the purpose of aligning the 'real world' with the absolute mission of jihad, refocus the public's attention on the United

States (Guantánamo in this case and the orange jumpsuits in which prisoners were kept) also served to invoke the more primordial thought of *tawhid,* of an absolute oneness, an all-encompassing truth that, as Derrida observed, finds its ultimate and unspeakable iteration in the Abrahamic 'gift of death'.[49]

The all-encompassing thought of *tawhid* was similarly invoked by the group's signature decal of a globe, as well as the ubiquitous white oval at the centre of the black and white flag of the *shahada.* Such symbolism prevailed over and against efforts to limit the group's public communications, as was the case with Abu Ayadh's banned interview with *Mosaique FM.* As the Saudi Arabian network *al-Arabiya* reported at the time, the decision to cut the broadcast mid-segment was predicated on a fear that Abu Ayadh would use the airtime to disseminate 'coded symbols and signals' that could compromise ongoing investigations or disturb 'public order' (*al-nizam al-'amm*).[50]

The Informatics of Terror

The perceived communicative threat surrounding Abu Ayadh's discourse was laid bare by the geopolitical breadth of the counter-communications campaign levelled against him and his movement. Of particular note in this regard were the WikiLeaks' revelations that Abu Ayadh had informed on a number of his former accomplices in the TCG, namely Abu Bilal al-Tunisi. The information was a minor footnote to the avalanche of headline material unleashed by the so-called Guantánamo Files. However, it was a detail that reinforced the ostensible objective behind one of the most complex communications operations of the post-revolutionary context.

In April 2011, WikiLeaks began releasing a series of secret Detainee Assessment Briefs from a stolen cache of US 'NOFORN' documents known as the Guantánamo Files. In four different briefs, all of which were published sometime between April 2011 and September 2012, Abu Ayadh appears as a government informant, his disclosures presumably conveyed amidst his period of detention in Mornaguia Prison where it was all but certain he had faced torture. The most stunning revelation contained in the briefs was from a memorandum on 4 November 2007, in which Guantánamo authorities detail the arrest history and threat assessment for Muhammad Ibn Arfhan Shahin, a Tunisian prisoner known by the *nom de guerre* 'Abu Bilal al-Tunisi'. Among other things the memorandum indicates that Abu Bilal had been 'a senior member of the Global Jihadist Support Network (GJSN) with specific membership in the Tunisian Combat Group (TCG) and the Armed Islamic Group (GIA)'. It quotes 'TCG founder, Abu Ayadh aka (Seifullah Omer Bin Hussein) [sic]' as having 'informed the Tunisian Ministry of Interior' that Abu Bilal al-Tunisi was 'on the TCG's advisory council and was responsible for TCG communications as well as welcoming and sending out new recruits'. In December 2014, Abu Bilal al-Tunisi was transferred to Kyzylorda, the birthplace of the Seljuk Empire and a former Russian military outpost in the south of Kazakhstan.[51] The date of his reported release, 30 December, was significant as a week prior to this a Jordanian military airplane had been shot down and its pilot taken prisoner by Islamic State fighters outside Raqqa, Syria. Shortly thereafter the pilot, Muath al-Kasasbeh, was burned alive inside a cage (on 3 January). The eventual retaliation, claimed by the Syrian army and advertised on its Facebook page, was said to include the assassination of 'Abu Bilal al-Tunisi' a 'leader of the Daʿish organization'.[52] The image tagged to the name 'Abu Bilal al-Tunisi' by *Al-Arabiya*, *RT* and other news outlets who reported on the strike did not match that of the Guantánamo detainee, who (more significantly) was reportedly still in US custody at the time of the pilot's capture. That two Tunisian mujahidin fighters would share the same *nom de guerre* was uncommon but not impossible. In the context of the black box of information that was the eastern Syrian warzone of early 2015, however, accuracy was optional. An alternative explanation for the Syrian (and Russian) assassination report was that it echoed and ultimately reinforced an existent narrative of US and UK involvement in the global jihadist phenomenon.

As one of Wikileak's key partners *The Daily Telegraph* of London reported at the time that the Guantánamo Files promised to 'expose' 'the crucial role that Britain has played in the global terrorist network that has been documented by those held at Guantánamo – with London emerging as a key "crucible" where extremists from around the world are radicalised and sent to fight jihad.'[53] Subsequent articles with headlines like: 'Terrorists radicalised in London', 'BBC part of "possible propaganda media network"', 'British aid budget funded key al-Qaeda aide', 'one quarter of freed Guantánamo Bay terror suspects "join insurgent operations"', and 'Julian Assange given peace prize' all aligned to foster the narrative that the US and its allies were complicit in the funding and promotion of jihad. It was a narrative that would carry all the way to the White House with the election of Donald J. Trump who famously repeated the position of Russian state-media that Hilary Clinton and Obama were the 'founders of ISIS'.[54]

Under the familiar refrain of 'secrets revealed', the WikiLeaks/*Telegraph* collaboration created a discursive framework in which jihadist actors like Abu Ayadh were not so much discredited as they were mobilised to demonise the greater hegemonic apparatus of the US War on Terror. Other news organisations that reported on the leaks, including *Der Spiegel*, *The New York Times* and *The Washington Post* pursued different lines of inquiry. While most took note of the revelations that many of the detainees were being held on perfidious grounds, they also used the stolen briefs to report on previously unknown plots to target sites in the US and Europe. Such reporting effectively steered the revelations towards the under-whelming conclusion that the US had kept much of the War on Terror secret. More importantly, the leaks revealed the extent to which the US relied on tenuous information, some of which had been extracted from forced confessions as was likely the case with Abu Ayadh at the Mornaguia prison. None, however, channelled the communicative agenda of Assange's Russian-aligned WikiLeaks organisation as explicitly as *The Telegraph*.

Remarkably, details concerning the Tunisians held in Guantánamo as well as Abu Ayadh's role in informing on them appeared not to surface in Tunisian media until mid-September, 2012. And even then, those outlets reporting on the news, including the online sites *Kapitalis* and *Tunisvision*, were relatively obscure. Either none of the country's mainstream media organisations read the complete docket of information (which was indeed obscured beneath layers of US military officialese) or no one reporting on Abu Ayadh's gang was prepared to complicate the jihadists' own 'Guantánamo narrative' in which the disappearance and abuse of fellow mujahidin served as a vivid point of departure in the struggle for redemption and revenge. That narrative, vitriolic and laced with the kind of transnational identity politics that fuelled the most strident rebukes from the country's leadership, provided an important point of antagonism in the post-revolutionary narrative of securitisation and state building.

Less important was approximation of the 'truth'. Obfuscated by the inherent obscurantism of jihadist rhetoric itself, the discourse on terrorism in Tunisia in the post-revolutionary context epitomised the notion as Hans Gadamer once wrote that all understanding is a 'historically effected event'.[55] The terror invoked by the narrative of Abu Ayadh was distinct from that which was cast by American discourse on terrorism, for example. There was little discussion of civilisational conflict in coverage of Ansar al-Sharia. And one would be hard pressed to find examples of the sectarian-laden descriptives ('*takfir*' or 'kharajite') so prolific in Saudi Arabian and Gulf Arab discourse on terrorism. Rather, not unlike the countries of the Southern Cone whose experience of terror was at times equally violent as North Africa, the major antagonistic forces in Tunisian discourse on terrorism were corruption, political deceit, drug and arms profiteering and the dismantling of the state by fugitive campaigns for autonomous zones and safe havens. 'In Argentina there do not exist mafias as they are known in the U.S. or elsewhere', wrote Gabriel Levinas, a major public intellectual who led the investigative team for a victim's association in Buenos Aires in the late nineties. The mafias in Argentina, unlike those in the States, were seen as 'independent', yet, at the same time, aligned by the 'flowchart' (*la organigrama*) of a corrupt political culture. They share each others' bottom line, metamorphosing hand in hand as the singular expression of a common condition.[56]

In Tunisia, where the war against terrorism was intertwined with decades of police corruption, 'emergency laws' and political abuse, it was of little surprise that the emergence of the Salafi-jihadist phenomenon would be greeted with an air of scepticism. Paralleling the absolutism of the jihadists' own discourse, the authoritarianism of the Ben Ali years – like all police states – obfuscated the truth beneath the shadow of the 'whole truth'. As observers like former Presidential council and sociologist Aziz Krichen understood, extremism appears illusionary in such contexts. The infiltration of 'business and mafia interests' (*intérêts affairistes et mafieux*) entailed not only a 'parasitic excrescence' (*excroissance parasitaire*) between corrupt private interests and the government, but a certain 'willful suspension of disbelief' on the part of the public such that iterations of extremism appear endemic to an omnipotent system of governance that simply, and naturally, reaps what it sows.[57]

The case of Mokhtar Belmokthar in Algeria (discussed in Chapter 7) gives evidence to this dynamic as well. As part of the STRATFOR intelligence files exposed by WikiLeaks it was revealed that the AQIM commander – a former military officer – had been given a green light by authorities to conduct operations in the Sahel so long as he promised to avoid targeting government installations. Neither the information that the government was implicated in the activities of AQIM nor that Mokhtar Belmokhtar ultimately betrayed the agreement were seen as a surprise.

Conclusion

Mired in the throes of revolution – the explosion of perspectives and new 'truths' (symbolised perhaps most elegantly by the June 2013 opening of the regime's information labyrinth and the victorious hacking of the elaborate algorithms once used to monitor people's electronic communications)[58] – the vitriolic narrative of the jihadists' own propaganda, obscurantist and dogmatic, cut across the complexities of the moment to deliver for adherents and critics alike, a story of extremities and providence and, perhaps above all, a sense of limitations.

Engrained in the democratic experiment has always been the impulse to confine it, to return to what preceded the experience of freedom, if only to set its course in motion once more. For those who followed it closely, the Arab uprisings revealed this phenomenon in concrete fashion. The slackening of censorship laws, the technological expansion of social media, parliamentary elections – all that led up to the great outburst of freedom that was the taking of the streets in 2010/11, was quickly followed by rapid trajectories towards the logical extremes of freedom's end: the opening of the prisons, the disbanding of police, the taking of revenge.

It was from this matrix of social convulsion that Abu Ayadh and his gang emerged as an ideological force that was at once radically anti-institutional, yet pathologically rigid in its opposition to non-conformational expressions of behaviour. The name 'Ansar al-Sharia' means, simply, partisans of the law. And Bin Hussayn's *nom de guerre* refers to a proto-Sunni cleric (al-Fudayl Bin Iyadh) in the court of the Harun al-Rashid (d. 766) whose authority in the *hadith* helped to undermine the legitimacy of the Caliph. (Curiously, several alternative sources exist describing a parallel historical figure of the same name who gained fame in early ninth-century Baghdad as a thief and highway bandit before turning to a life of asceticism.)[59] In today's currency, the name also summons a curious blend of chaos and control, of the criminal and the correct. It is through his story, or the construction of his story, that one begins to *see* the Platonic paradox: 'that which democracy defines as good', namely, freedom, 'is also what destroys it'.[60]

Socrates' antithesis of democracy is not tyranny, but freedom, that for which democracy provides a politics, a narrative and an antidote. Abu Ayadh's project mirrored the project of democracy, providing a politics to the apolitical and authority to the anti-authoritarianism that best characterises the young cultural code of '*jihadi cool*'.[61] The phenomenon of Ansar al-Sharia posed a significant challenge to those in the centre. While the discourse on criminality has elsewhere served to reinforce an oppositional pole, positioning detractors as anarchistic, the rhetoric of Ansar al-Sharia was immune to such rationale as it stemmed from an ideology founded wholly upon the utopic idea of absolute control. It is for this reason, I would posit, that the discourse on *laïcité* in post-revolutionary Tunisia proved to be so precarious for those in positions of authority. The secularists (for want of a better word) found little recourse but to identify their political campaign within the narrative of 'Democracy', writ large, and, more explicitly, the narrative of Westernisation, both of which signalled a kind of chaos when held against the conservative palate of Salafism.

The story of Abu Ayadh, then, *l'homme revolté*, became at once revolutionary and counter-revolutionary. In a yet more literal translation of Camus's famous phrase, Ayadh himself, we might say, was a *man revolted*. A famed jihadist imprisoned for an uncertain crime, the revolution which released him soon thereafter indicted him in the court of public opinion. He was to become a symbol of radical freedom despite the very essence of his ideology. And his disappearance, cloaked in mystery, would serve only to reinforce the permanency of the paradox he represented.

4

Media Wars I: Egypt

Introduction

In the immediate months and years following the collapse of the old regimes the fields of conflict in the Arab world began to multiply. While Syria, Libya and Yemen spiralled into civil war, a less bloody though equally charged information war was raging within the capitals of Egypt and Tunisia. In its most extreme iteration this conflict manifest in a series of high-profile censorship cases, including the incarceration of writers and journalists. But a broad spectrum of ideological fault lines rippled through public discourse in more subtle ways as well. Much of this information war concerned questions of veracity, seemingly basic accounts of what had occurred, what was gained, and what was lost. A multiplicity of perspectives, unprecedented for countries whose public discourse had long been suspended in the grip of state-censorship, propelled new and old voices alike into the role of statesmen, historians, activists and oracles. As time progressed and the memories of protest calcified into stories, the role of narrative became ever more dominant, not simply as a manner of recollection but as a dividing line between new and emerging camps of identity, and, ultimately, power.

This phenomenon traversed the entirety of the so-called Arab Spring world, setting in motion acute though disparate resolutions. Bassam Haddad, an American-based academic and co-founder of the influential e-zine *Jadaliyya*, wrote of the war in Syria in the fall of 2016:

> There has been increasing gravitation toward two mutually exclusive narratives: (a) that of 'pure and consistent revolution', and (b) that of 'external conspiracy'. Both narratives carry grains of truth, but both are encumbered

by maximalist claims and fundamental blind spots that forfeit any common ground necessary for enduring cease-fires or potential transitions, as well as postwar reconciliation.[1]

In Egypt where all-out war was narrowly evaded, narratives became ever more explosive in part because they served to control through language what action might otherwise mitigate with force. 'After the fall of Mubarak', Alaa Abd El Fattah wrote, 'the state was forced to compromise with the revolution while trying to contain it by appropriating its story.'[2] This dynamic was particularly vivid in the heady months of public debate preceding and immediately following the drafting of a new constitution in the autumn of 2012. But there, as with the closing days of Tahrir, a familiar theme emerged from public debate: that the Muslim Brotherhood had eclipsed the aspirations of the revolution with the goal of creating an Islamist state and that the military, once again, should intervene for the sake of the nation's survival.

'*Eid wahda*', the dominant nationalist motif reincarnated from its previous iteration as a message of religious unity and deployed through state-media outlets in the early days of the uprising by people like Mohamed Hassanein Heikal, created a framework for the mobilisation of a vast social movement – 'Tamarrod' (Rebellion) – that would culminate in the ousting of the country's first elected president, Mohamed Morsi. It also provided the backdrop for the violent consolidation of the campaign as Egyptian state security forces moved to evacuate supporters of Morsi from two squares in a suburb of Cairo. Human Rights Watch estimated at least 817 people were killed during the Rabaa massacre.[3] As with the discourse on terrorism in Tunisia, the 'Tamarrod narrative', to coin Jean-Pierre Filiu's usage,[4] vectored towards the periphery of the country where the government pursued an open war on terror in the Sinai. It encompassed as well the many sporadic acts of violence that peppered the capital. Even attacks claimed by outside groups tended to be attributed to the Muslim Brotherhood.[5] But as with discourse surrounding the onset of protests in January 2011, the Tamarrod narrative was fuelled by a range of complex communications strategies, some of which originated beyond the shores of Egypt.

Incitement Aesthetics

In one of the most explosive such instances, the phrase *akhwanat al-dawla*, or the 'Brotherization of the State', emerged in dialectical fashion with the sudden distribution of the bizarre online film *The Innocence of the Muslims*. Distributed via YouTube in the summer of 2012, the film, which depicts the Prophet Muhammad as a lewd and violent misogynist, was subsequently subtitled and exposed to Egyptian audiences just a week before the historically charged tenth anniversary of the 11 September attacks.[6] Popular response to news of the film was swift and US embassies throughout the region were besieged, including in Egypt, where protesters raised above the wall of the compound a black and white flag inscribed with the *shahada* – the soon-to-be infamous trade mark of the Islamic State group In Tunis, the US embassy was attacked along with an adjacent American school. And in Libya, in the eastern city of Benghazi, the US Ambassador Christopher Stevens, along with three other Americans, was killed in a violent assault on the compound where he was sheltering.

The channels through which the film was made public remain obscure. *The New York Times* reported on the efforts of a Coptic American with ties to the then unknown director of the movie, Nakoula Basseley Nakoula, to disseminate the film abroad. Seemingly in coordination with the extremist American preacher Terri Jones, who's 'International Judge Muhammad Day' (which involved burning copies of the Quran) was scheduled for 11 September, Morris Sadek, the American Copt, published links to the Arabic-subtitled video along with a derogative cartoon on his blog.[7] News about the video flamed through social media and then entered the mainstream during a 8 September interview on *Al-Nas TV* (a Saudi-owned religious satellite channel) between the host of the programme *Misr al-Jadida*, Sheikh Khaled Abdallah and Muhammad Hamdi Omar, a self-described 'researcher in religious doctrine'. Omar said he had 'discovered' the video (referred to only as *al-film al-Musi' l-il-rasul*) the previous day on YouTube.[8] On 9 September, in an ostensible attempt to get ahead of the backlash, the prominent Coptic lawyer Mamadouh Ramzi, speaking on *Al-Arabiya*, emphasised that his community was against any film that could be seen as 'offensive towards the Prophet Muhammad' and recalled that Copts were in a defensive position following the massacre at the foot of the Maspero offices along the corniche the previous year.[9] Ramzi's comments, however, did little to stem the popular outcry. In an apparent last-ditch effort to divert anger from the Coptic community, Noukala told the *Wall Street Journal* that his name was 'Sam Bacile' and that he was a Jewish-American with ties to Israel.[10]

By the morning of 11 September, the impact of the video and the informatics campaign surrounding it was a fait accompli. Broad swathes of Arab society turned out in protest. Alongside more radical Salafist activists, secular groups, including Egyptian soccer supporters known as 'Ultras' rallied at the American embassy to protest. As Amru al-Ansari recalled in his 2014 study of the group, the Ultras largely stayed away from the Tamarrod campaign, a remnant in part of the more visceral reaction shared in response to the 11 September anniversary events. None of their primary Facebook accounts, for example, showed endorsement or links to the penultimate call for protest against the Morsi government on 30 June.[11]

The video and the anniversary created a perfect storm of misinformation, ideological positioning and narrative deceit. Citing the video, the governments of Pakistan, Bangladesh and Sudan blocked access to YouTube. Iran shut down access to Google (the parent-company) entirely. Russia, which covered the event extensively through its flagship outlets *RT* and *Sputnik* indicated that it might also ban YouTube in response to the video. Appearing conciliatory to a Muslim population that it seldom defended, Russia's response, on the surface, seemed an unlikely one. As part of the country's bitter grievance with Silicon Valley over the perceived role of social media in the so-called Snow Revolution of 2011–12, however, the incident provided an opportune moment to challenge the media giant Google on more clearly defined, if politically charged ethical grounds.[12]

In response to the video, and with little regard to the violent assaults on the US and Israeli embassies, Brotherhood lawmakers in Egypt (like in Tunisia) doubled-down on the creation of new anti-blasphemy laws. Although the proposal had been in motion for months prior to 11 September, the video controversy and the protests that followed appeared now to illustrate the extent to which Brotherhood policies were driven by extra-national interests and extra-official actors. It was a narrative that was becoming increasingly vivid on social media where an alternative, if bizarre theory behind the video controversy had begun to take hold.

The Brotherisation Backlash

Citing a WikiLeaks document delivered to *The Daily Telegraph* of London amidst the opening salvos of unrest in January 2011, writers Daniel Greenfield and Frank Gaffney of the extreme right-wing American news blog *Front Page Magazine* used the attack on the embassies and the death of Ambassador Stevens to advance the myth that the Obama administration was conspiring with Islamists, including the Muslim Brotherhood, to alter the political landscape of the Arab world. The death of Ambassador Stevens, they held, was a result of perilous naiveté, a case of 'feeding the crocodile'.[13] The WikiLeaks document the authors refer to was written in 2008 and stolen in 2009. *The Telegraph* ultimately published the cable under the headline 'Die Hard in Derna'. A typical kind of diplomatic dispatch, the cable indicated that Stevens had developed contacts in the Libyan coastal city and that his interest in the struggling economy spurred willingness to see past the prevailing stereotype of Derna as a haven for extremism.[14] As spun by *Front Page Magazine*, however, the document served to indicate complacency towards extremism if not complicity on the part of Stevens and the Obama administration with the region's growing wave of Islamisation. Islamists and the Muslim Brotherhood had 'infiltrated' the Obama administration, they wrote, and it was for this reason that the US had intervened to prevent Ghaddhafi from invading the province of Cyranica, including the city of Derna, where the group thought to be responsible for the assault on the US compound and embassies, Ansar al-Sharia, maintained a stronghold.[15]

Appearing just two days after the ambassador's death, Greenfield's article gained traction among several far-right circles in the US that had been pushing a racially imbued attack on Obama as sympathetic to Islam more or less since his inauguration in 2008.[16] In Egypt, however, with the official launch of the Tamarrod campaign still months away, the narrative of US complicity in the Brotherisation of the state intensified. This, despite the ostensible contradictions of the 11 September video conspiracy.

Significantly, some of the most virulent proponents of the Brotherisation narrative emerged from the recently created satellite stations *Tahrir News*, co-founded by Ibrahim Eissa, and *ONTV*, the popular satellite channel started by the Coptic-Egyptian billionaire Naguib Sawaris. Tahani al-Jabali, the former Vice-President to the Supreme Constitutional Court and a regular commentator on *ONTV* became one of the narrative's most identifiable voices. In an extended interview from 31 December 2012, she framed the phenomenon of Brotherisation in black and white terms. 'We have revolted against the legacy of Gamal Abdel Nasser', she said in reference to the election of the Muslim Brotherhood. 'Yet we considered him still the greatest source of our modern existence. Gamal Abdel Nasser said any revolution that undermines the youth does not deserve the honor of being called a revolution.'[17] The origin of this betrayal, she held, was a conspiracy. Or as one popular segment was inconspicuously titled: 'The American relationship with the Muslim Brotherhood'.[18] Amr Ammar, a regular contributor to *Tahrir News*, reproduced verbatim Daniel Greenfield's *Front Page Magazine* article on Ambassador Steven's death in his 2014 book *Al-Ihtilal al-madani* (The National Humiliation). A towering monument to post-revolutionary conspiracy theories, Ammar's book, which received broad attention in Egyptian media, was dense with articles translated from explicitly anti-Muslim websites like 'ACT for American Houston', a proto fake-news site that became active on the part of Donald Trump during the 2016 US elections.[19]

US officials were markedly unresponsive, if not unaware of the communicative assault being levelled against the Morsi government at their expense. In June 2013, the American Ambassador Anne Patterson held a high profile meeting with the soon-to-be disgraced deputy of the Muslim Brotherhood Khairat al-Shater and the Supreme Guide of the Muslim Brotherhood Mohamed Badie. Staged in the midst of a media firestorm, *Youm 7* and other mainstream media outlets spun the meeting as proof of US complicity in the Brotherisation conspiracy. A subsequent 18 June speech in which Patterson discouraged Egyptians to take to the streets was met with fury. Michele Dunne, a former State Department official and specialist in Egyptian politics who was herself the subject of an aggressive misinformation campaign in 2012, recalled how Patterson was shouted down by the audience in the midst of her speech at the Ibn Khaldun Center in Cairo. The Egyptian government, she noted, had previously employed tactics of 'pre-emptive intimidation' with the aim of making diplomats question whether they should speak out in public.[20] But the full-throated attack on Patterson, including a Trump-like 'lock her up' visual media campaign, illustrated just how belicose the Brotherisation strategy had become.

Another robust engine of influence behind US–Brotherhood conspiracy theories was *RT* Arabic. Ashraf al-Sabbagh, a Russian-educated journalist and translator, surmised the imagined relationship succinctly in his 20 June 2013 column for the Russian state outlet:

> There is no doubt that the Brotherhood provides all the necessary guarantees for the security of Israel. Even if these guarantees come at the expense of Egyptian security in the medium and long term. The equation (Qatar – Hamas – Brotherhood) gains clarity when held against the general scenario of Islamizing the region. Imprinting a religious character [on the Middle East] provides Israel not only security, but also the rights to its full Jewishness.[21]

Gleaning a series of seemingly incompatible nodes (Israel's open hostility to the Brotherhood party of Hamas most notably), al-Sabbagh's report illustrated the degree to which narrative conceit not only thrives in the absence of clarity, but requires as much. Evidenced here by a veritable swirl of information (the report goes on to explain US opposition to Assad and a certain relationship with Iran which al-Sabbagh describes as 'too complicated' to unpack), the narrative behind *RT*'s myriad forays into the Brotherisation debate epitomised what Peter Pomerantsev described as the 'hook' and 'distract' strategy of Kremlin state media outlets; the objective being not to provide a 'counternarrative' *per se* but rather, simply, to 'disrupt Western narratives' in the most effective way possible.[22]

Needless to say, *RT* (while increasingly present in the Arab world) was hardly the most pervasive source of information in Egypt during the presidency of Mohamed Morsi. But its engagement with the Brotherisation narrative was indicative of the degree to which the country under Morsi had become a magnet for ideologically contentious narratives from inside and outside its borders. Russia, for example, had targeted the Muslim Brotherhood as a designated terrorist organisation a decade before the label was adopted by countries like Syria (2013), Egypt (2013), Saudi Arabia (2014), Bahrain (2014) and the United Arab Emirates (2014).[23] Its campaign against *al-Jazeera* and Qatar was similarly well established. In early December 2010, WikiLeaks sent to *The Guardian* a stolen State Department cable with the headline 'Cables claim al-Jazeera changed coverage to suit Qatari foreign policy'.[24] Generously spun from an informal communique in which the former US ambassador to Qatar describes the political viability of the news organisation to the Qatari government, the information became a convenient focus for a viral onslaught of politically-tinged critiques, including *al-Jazeera*'s satellite dominance, the country's hosting of the 2022 World Cup, its hosting of US Central Command at al-Udeid Air Base, and the increased presence of political Islam (welcomed within Qatar) on the world stage more generally.[25]

More pointed than the Qatari dimension of the Brotherisation narrative was the virulent inculcation of the Obama administration. From the President's birthright, to Obama's appointment of the special counsel to the Muslim world, Rashad Hussein, or Secretary Clinton's choice of personal assistants in Huma Abedin, the narrative of Brotherisation was rife with theories that the US government was sympathetic to Islamists and secretly favoured religio-political establishments in the Middle East more generally, be it the Muslim Brotherhood in Egypt or the Jews in Israel. Its strength was fuelled by social media and somewhat, ironically, by Americans themselves.

It is worth revisiting in this regard the genealogy behind the rumour that Obama and Clinton were the 'founders' of ISIS which was an extreme iteration of the same narrative pattern.[26] The precise origins of this informatics campaign, which created a stir in the context of the 2016 American presidential campaign, remain obscure. But as with the narrative of Brotherisation, its genealogy suggests that the influence operatives behind it intended the impact primarily for Middle East audiences. In the US, debate behind the preposterous claim, emphatically repeated by Donald Trump, typically circled around banal semantic distinctions. '"Facilitat[ed]" and "found[ed]" are very different words' tweeted the American Ambassador to Russia Michael McFaul in response to the argument by Trump defenders that he was speaking 'metaphorically'.[27] In the context of the Arab media sphere, however, the narrative calcified around an existent and recognisable motif of Western imperialism. As early as August 2014, just a month after Abu Bakar al-Baghdadi's declaration from the pulpit of the Mosque of al-Nuri in Mosul, major Arab news outlets, including the Egyptian newspaper *Youm 7*, were quoting right-wing US conspiracy blogs like *Veterans Today* and circulating 'reports' that Edward Snowden had disclosed the true identity of Abu Bakr al-Baghdadi as 'Elliot Shimon', an Israeli Mossad agent.[28] The rumour – which appeared in early July with online sites like *Arabesque* (8 July 2014) falsely citing *The Intercept*, or *RT* (17 July 2014) citing *Gulf Daily News* – stemmed from Iranian state news services. On 18 June, just days after al-Baghdadi's speech, *Fars News Agency* quoted Iran's Chief of Staff of the Armed Forces, Major General Hassan Firoozabadi, as saying 'ISIS "is an Israel[i] and America[n] movement for the creation of a secure border for the Zionists against the forces of resistance in the region".'[29] With the effective reporting of people like Aryn Baker for *Time* magazine, the rumour failed to gain traction prior to Donald Trump. In much of the Middle East, however, the story never lost its lustre.

Morsi and the Brotherhood at times reinforced core aspects of the Brotherisation narrative. Alongside a spike in gas prices, the collapse in health and food services, and lapses in security, the leadership's rhetorical emphasis on transnational issues, including Egyptian-Israeli relations and Shiaa-Sunni tensions in Iraq, fed into the notion that the organisation was indifferent, if not hostile, to the project of nationalism.[30] The 2012 Draft Constitution compounded the problem. Among the most controversial components of that document were the proposed judicial independence of Al-Azhar in matters pertaining to Islamic law (Article 4) and the implementation of a blasphemy code (Article 44) prohibiting 'insult and abuse of all religious messengers and prophets'. Attempting to quell public outcry, Morsi launched a live national broadcast defending the document by way of his group's democratic legitimacy, reminding audiences that the Brotherhood was in power as the result of two previous elections (19 March and 20 June 2012). In a nod to the pre-revolutionary motif of *eid wahda* he exclaimed that his party's supporters included 'men, women, Christians, and Muslims'.[31] In light of the ascendant, post-revolutionary version of *eid wahda*, however, his argument did little to appease critics who viewed the supranational dimensions of the Brotherhood and its perceived ties with Washington as inexorably antagonistic to the nationalist ethos of Tamarrod. As Mustafa Bakri wrote in his official biography of al-Sisi: 'the army was convinced that [the chaos] occurring [in early 2013] was a plot by the Brotherhood with direct support of the Americans'.[32]

The same antagonistic front that had emerged to eclipse the initial occupation of Tahrir was once again driving the course of events.

Censoring Dissent

Fuelled by the Brotherisation conspiracy, the Tamarrod campaign infiltrated and ultimately subsumed an otherwise eclectic field of post-revolutionary discourse. An early, illustrative such instance occurred in February 2014 when the essayist and blogger Belal Fadl made international headlines for a satirical article that was reportedly 'banned' from publication in *Al-Shorouk*.[33] 'The political marshal' begins with the explosive assertion that Mohamed Hassanein Heikal was helping to prepare the presidential campaign of then Defense Minister Abd al-Fattah al-Sisi. Its subsequent content – draped in an allegorical accounting of a conversation between Heikal and the British military commander Bernard Montgomery on the occasion of the twenty-fifth anniversary of the Battle of El Alamein – levelled a stringent critique of al-Sisi's imminent presidential bid as well as his seemingly spontaneous promotion from General to Field Marshal. But it also took aim at the narrative, and indeed, the principal narrator of *eid wahda*, the ideological paradigm from which the Tamarrod narrative was born.

Despite Fadl's explanation that he was at once 'proud to contribute to the overthrow of the Brotherhood' *and* concerned about the 'militarization of the country and the return of state repression', the shadow of *eid wahda* had begun already to eclipse any inkling of public ambiguity.[34] In the face of such a dominant ideological structure there was to be no room for 'and'. One's support for the regime was a matter of 'yes' or 'no'.[35]

While the controversy surrounding the Fadl article appeared to be contained – the author resigned from his position at *Al-Shorouk* and issued an apology – the editors' act of self-imposed censorship set the stage for a cascade of similar incidents.

On 2 June 2014, the political satirist Bassam Youssef cancelled his self-styled *Daily Show* – *Al-Bernameg* – indicating the station broadcasting the programme, *MBC Misr*, had faced pressure from authorities. Or as he said in his closing broadcast: 'the media climate' was prohibitive to his kind of humour.[36] Later that same month, the novelist and vocal advocate of the 25 January uprising Alaa al-Aswany resigned from his column with *Al-Masri al-*

Youm writing: 'nothing is allowed but one opinion and one thought and the same words. Criticism is no longer allowed, neither are divergent opinions. Nothing is tolerated but acclaim at the expense of the truth.'[37] High profile resignations and cancellations continued throughout the year. In October, the popular news show *Al-'Ashira masaian'*, hosted by Wael Elebrashy, was abruptly pulled off air as the host began to report on an itinerant woman forced to give birth on the steps of Kafir al-Dawar hospital.[38] The show's producer had reportedly been told by government officials to monitor Elebrashy's reporting after a series of stories on the death of a child in a Cairo public school prompted the Ministers of Education and Housing and Urban Communities to threaten resignation.[39] Writers and journalists outside the public spotlight faced more direct forms of censorship. In 2014, the poet Fatima Naoot posted a note on Facebook criticising the slaughter of animals for Eid al-Adha. Two months later, a public prosecutor referred her for trial citing contempt of religion, a crime in Egypt. She was sentenced to three years in prison and fined 20,000 EGP.[40] Her case was ultimately suspended upon appeal. Others, like the writer Ahmed Naji, imprisoned following an obscenity charge for his 2014 novel *Istikhdam al-haya* (The Use of Life); Ismail Alexandrani, who was sentenced to five years for reporting in the Sinai; photographer Mahmoud Abu Zeid, who was imprisoned for documenting clashes between Morsi supporters and the police. or blogger Alaa Abd El Fattah, who was sentenced to five years for inciting protest, were not as fortunate.

Egypt's war on words was hardly unprecedented. Samir Farid, an Egyptian film historian, observes in his 2002 book *Tarikh al-raqaba 'ala al-sinima fi Misr* (The History of Censorship in Egypt) that the first formal censorsing of the press was a 1904 British law (following a decree from 1881) that had been designed to stem a surge in critical reporting surrounding the two-year-long Orabi revolt (1879–82).[41] Efforts to block the Egyptian press stretch back even further if one considers the Ottoman Printing and Publication Law of 1857, which, as Ami Ayalon notes, first attempted to regulate the press in Egypt through licensure.[42] But free expression and the move to counter it are something of the brick and mortar to Egyptian authoritarianism. As Khalid Kishtany wrote in his lucid 1985 book *Arab Political Humor*, it was the pioneering prose of satirists in the pages of *Rose al-Youssef*, one of the country's most powerful vehicles of dissent since its founding in 1925, that prompted a 1933 law prohibiting the 'treatment of journalists as political prisoners in any offence affecting any member of the royal family'.[43] The magazine had lambasted Egyptian officials for their acquiescence to British officials, noting the eagerness of the country's leaders to convene their submission to the colonial power.[44] Rather than be detoured by the law, satirists at *Rose al-Youssef* broadened the scope of their critique to include the country's wealthy aristocracy and even the liberal opposition party al-Wafd. Kisthany highlights for example a set of cartoons depicting the Anglo-Egyptian Treaty of 1936. One especially telling image shows the British representative John Bull sitting beside the leader of the Wafd's Mustafa al-Nahhas Basha and 'al-Masri Effendi' the magazine's stock 'everyman'. Nahhas Pasha tells al-Masri Effendi to 'pick up a snake' (the treaty) lying on the table. '"It's true its bite is painful and fatal," he tells him, "but this is the smallest snake I could find you."'[45]

In 1947, the British-backed government of King Faruq passed a series of directives in response to a string of films during the war years about the struggle for labour rights. All produced by the newly created Studio Misr, films like *al-Suq al-sawda'* (Black Market, Kamal al-Tulamsani, 1943/47), *al-'Amil* (The Worker, Ahmad Kamal Mursi, 1943), and *al- muzahrat* (Manifestations, Kamal Salim, 1941) tapped into the political power of realism, a genre first introduced to Egyptian audiences in 1939 with Kamal Salim's *al-'Azima* (The Will).[46] The 1947 directives in turn outlawed any film that inspired or contained 'communist propaganda against the Royalty, or the existing system of government, or social probity'.[47] It forbade films that displayed 'social revolutions, demonstrations, or squalidness'; 'dirt lanes'; the 'homes of poor peasants'; 'workers on strike'; or 'attacks' by workers against their employers and 'vice versa'. Significantly, it also targeted images 'of crime in favor of workers', or films that 'spread the spirit of rebellion among them as a means to claim their rights'.[48]

Following a period of neo-liberal reform and a boom in satellite television production, the back and forth between free expression and censorship intensified with the advent of digital communications technology in the early twenty-first century. In 2007, Reporters without Borders noted that the Egyptian government had 'reinforced surveillance of the Internet in the name of the fight against terrorism', creating a 'special department within the Ministry of Interior'.[49] The move appeared in response to a marked increase in civilian journalism, including the work of bloggers following the contentious re-election of Mubarak in 2005. 'We bloggers decided to take matters into our own hands', wrote Wael Abbas in *The Washington Post*.[50] 'I took photos and video footage of the demonstrations and posted them on the Internet, restricting my comments to simple explanations of what was in the pictures.' Abbas wrote that he was receiving 30,000 visits to his page a month when he recorded and uploaded a scene of police brutality on the day of Mubarak's reelection in May 2005. 'My site received half a million hits in two days', he recalls. 'It caused a huge scandal for the government.'[51]

Tensions between the country's chattering class and government censors reached new heights following the unprecedented opening for free expression created by the 2011 uprising. But by late 2016, the government of al-Sisi had passed legislation that allowed, among other things, the revocation of media licenses by a new, all-powerful regulatory body: the Supreme Council for the Administration of the Media.[52] Several months later, the government blocked access to a score of critical international websites and news organisations, including *Al-Jazeera*, *The Huffington Post* in Arabic and *Mada Masr*, one of the country's last domestic bastions of critical dissent.[53]

International outcry followed a number of these cases with much of the criticism abroad landing on the shoulders of the country's newly empowered President, Abd al-Fattah al-Sisi. But the narrative lines of critique underpinning the crackdown existed well-before al-Sisi's coming to office.

Civic Narratives

The rise of the Tamarrod narrative, it should be said, was not an inevitability. In the months following Mubarak's resignation, from March to July of 2011, competing factions capitalised on the ostensible vacuum of authority flooding the airwaves of nightly talk shows, radio programmes, news columns and blogsites with a veritable deluge of information. State security clearances previously licensed by the General Authority for Investment, an arm of Mubarak's innermost circle, were no longer required.[54] The Ministry of Information was briefly abolished.[55] Sixteen new satellite channels appeared at this time,[56] a surge that represented 'about a 30 percent increase' in Egyptian broadcasting.[57] The Muslim Brotherhood launched *Masr 25*. Mohamed al-Amin Ragab, a businessman with ties to the Mubarak regime, started the Capital

Broadcast Center (CBC) which featured, in Abdalla F. Hassan's words, a 'bouquet of channels'.[58] Naguib Sawiris, the billionaire telecommunications and construction investor who was reported to have fled Egypt when the uprising began, sponsored two new channels.[59] The 'maverick journalist' Ibrahim Eissa, who appeared arm-in-arm with Mohamed ElBaradei at the outset of the uprising, started, along with Ibrahim al-Muallim, the founder of *Al-Shorouk*, the newspaper *Al-Tahrir*. At the same time, the interim authorities, led by the Supreme Council of Armed Forces, sought to 'retool' programming out of the Egyptian Radio and Television Union.[60] It was this body, created in 1979 and revised in 1989, that laid the original groundwork for the overarching authority of the Ministry of Information.

Amidst such proliferation of voices the scramble for airtime became not simply an assertion of free expression but a race to write history. 'On January 28 what happened happened', wrote the influential journalist and public opposition figure Hamdi Qandil. 'The leadership of the brotherhood broke from prison, with or without the help of Hamas, and once it had become clear to them that the police forces had collapsed they decided to join the revolution.'[61] Qandil, whose long career in television stemmed from the Nasserite years when he became the nation's first nightly news broadcaster in 1962 with the programme *'Aqwal al-suhuf'* had had his own experiences with censorship, including the abrupt cancellation of his programme *'Ra'is al-tahrir'*.[62] As public discourse surrounding the future of Egypt intensified in the lead up to the country's first open elections in May 2012 it was familiar voices like that of Qandil that served to provide a degree of stability. By framing events of the recent past within a clearly delineated ideological paradigm of nationalist thought, it was also this spectrum of opinion that would outlast all subsequent voices of dissent to reestablish the country's dominant narrative identity. The relative centrality of this position became all the more important as Egypt's myriad opposition parties sought tenuous new lines of affiliation, most of which would collapse almost as soon as they were established.

On the night of 3 March 2011, just weeks after Mubarak's resignation, Qandil found himself along with Alaa al-Aswany, Naguib Sawaris and Ahmed Shafik – all well-known public figures – in the midst of a media spectacle when al-Aswany tore into Shafik, the acting Prime Minister, for his past relationship with the Mubarak regime. Broadcast live on the popular evening talk show '*Baladna bil-Masri*' hosted by Yousry Fouda, a major proponent of the 25 January uprising who would become himself a target of the Tamarrod campaign,[63] the exchange sent immediate ripples through the secular political establishment. The following day, Shafik abruptly resigned from his post as Prime Minister, a development hailed as a triumph of free debate for critics of the regime. But the move would prove more insidious than was first apparent. Shafik, as it turned out, was merely retooling for his presidential bid the following spring, and al-Aswany, like Fauda, soon found himself on the receiving end of proto-nationalist attack campaigns, most often for his association with Muhammad ElBaradei. While subtle, the reversal was a powerful one. As Qandil recalled three years later:

> I kept quiet during the outburst for two reasons: the first was to counterbalance Al-Aswany's sweeping onslaught. The second reason was not known at the time. I had met with Shafik in his office, just after he had been commissioned to form his ministry, and I had made my remarks clear to him at that time.[64]

Along with al-Aswany and Abd al-Jalil Mustafa, Qandil had been part of a group he refered to as the 'Automobile Club'. A gathering of reform oriented intellectuals, members of the group helped spearhead what became known as the 'Fairmont Accord', an agreement between the Muslim Brotherhood and different opposition factions – a consortium known as the 'The Committee of a Hundred' – dedicated to honouring the ideals of Tahrir.[65] The accord, which included people like Wael Ghonim and Ahmed Maher from the April 6 Movement as well as older opposition figures like Qandil, Abd al-Jalil Mustafa and al-Aswany, had been created in an effort to leverage moderation in the face of an imminent Muslim Brotherhood victory. 'We did not trust them because they were not a revolutionary group', Mustafa said of the Brotherhood.[66] As evidenced on the stage of *ONTV*, however, the Automobile Club along with the Youth Coalition movement, were equally, if not more concerned about the prospect of a Shafik presidency. Ahmed Maher, the founder of the April 6 group said later: 'we were left with no choice . . . Shafik's success would have symbolized the end of the revolution.'[67] For Morsi and the Brotherhood, the Fairmont Accord, announced just two days prior to the final run-off vote, provided a path to broad legitimacy, a symbol that the 'national forces' were in support of their campaign.[68] Capitalising on the agreement, Morsi exclaimed the election of Shafik would be like 'aborting the revolution'.[69] For the Automobile Club and the Youth Coalition the agreement appeared to signal an almost absurd ideological concession. Few ideological platforms were less compatible with that of the Muslim Brotherhood. The position assumed reason within the scope of the myth of succession (see Chapter 2). But what Morsi and the Brotherhood appeared not to accept was that the myth of succession served merely as a complementary node in the inevitable ascendance of *eid wahda*.

Any party on the opposite end of the nationalist narrative was bound for trouble. Ahmed Shafik evinced full awareness of this: 'don't play the nationalist card with me!' he shouted in response to al-Aswany's demand he assume accountability for those injured during the protests.[70] Leveraging the nationalist narrative became tantamount to political survival.

Fourth Generationalism

At the heart of Egypt's post-revolutionary information war was a resilient pre-occupation with the power of information itself. The Muslim Brotherhood represented in this discourse not simply a post-secular challenge to Egyptian unity but, as Hamdi Qandil alluded to, a veritable threat to the very borders of the nation. Such was the explicit message of the forty-eight hour ultimatum issued by the Armed Forces that capped the Tamarrod campaign on 1 July 2013.

> Egypt (*al-saha al-misriya*) and the whole world witnessed yesterday the demonstrations by the Egyptian people and their unprecedented show of peaceful and civilized expression. Everyone saw the movement of the Egyptian people and they listened to its call with the utmost respect and attention. It is imperative that this movement be responded to by all parties who carry the burden of responsibility amidst these dangerous circumstances surrounding the homeland.[71]

Disseminated live on air and through the military's various social media sites, the proclamation accompanied a slickly polished photograph of then Defense Minister Abd al-Fattah al-Sisi. Staring directly ahead and stripped of affect, the General's image along with the disembodied voice overlay accompanying it invoked the apparition of an automaton. Along with the ominous threat of a forty-eight hour ultimatum, this dramatic entrance into the centre of public discord accompanied a broad movement on the part of the military to reconstitute the narrative of the revolution within a narrower, more definitive set of terms. The Egyptian arena – or physical territory (*al-saha al-misriya*) – had been besieged by a dangerous set of circumstances. The threat was not simply ideological but material, menacing industry and financial systems, social networks, media and, above all minds. It was a threat that had animated the core of the nation – the 'Egyptian people' – and consolidated the will of the republic to fight back through the expression of the armed forces. The Muslim Brotherhood with its international ties and extra-territorial rhetoric coalesced in symbolic terms with the more sophisticated threat of 'Fourth Generation Warfare' – an idea and a phrase that the future President al-Sisi would define in unmistakably populist terms as 'the devil's plot' (*al-mukhatat al-shaytani*).[72]

Fourth Generation Warfare, which intersected in official discourse with the transnational ambitions of the Muslim Brotherhood, asymmetrical terrorist operations, the techno-revolutionary knowhow of the April 6 group, *Al-Jazeera*, and the machinations of US and Israeli spy agencies, spoke not simply to the fear of territorial invasion but to the imminent prospect of a future collapse. As early as 1992, as Laurie Brand observed, the Egyptian military had begun to express a prioritisation of 'treating knowledge metaphorically as the weapon of the future'.[73] And it was precisely in such tones that al-Sisi and the military positioned its coming assault on the social and intellectual legacy of the 25 January uprising.

As revealed in a cache of documents ostensibly leaked from the Ministry of Interior in the spring of 2014, the government's strategy for waging its counter-offensive included a full-scale mobilisation of the country's ICT surveillance infrastructure. Unlike Tunisia, where legislators famously moved to 'neuter' the Tunisian Telecommunications Agency, the Egyptian government had retained much of its pre-uprising footing.[74] Highlighting a strategy of searching for keywords and terms deemed 'threatening to the law and public morality' (*al-qanun wa al-adab al-'amm*), the government's DPI, or Deep Packet Inspection technology, appeared to be operating as it did before the uprising. However, the 'leak', which was published in the staunchly pro-Sisi newspaper *al-Watan*, also emphasised the Ministry's use of its 'iron grip' technology for combating the 'security risks' posed by 'social networks'.[75] As much a spotlight on the government's existent capabilities, the information functioned as a propaganda device in the government's increasingly populist campaign for 'law and order'.

Underscoring the government's material tactics in combating online threats there emerged a range of sympathetic cultural productions from non-state and state-actors alike. In late November 2013, the Egyptian state satellite channel *Al-Oula 1* ran its first episode of a documentary series, *Hurub al-jil al-rabi'*, or Fourth Generation Warfare. By 2015, the series had expanded from a weekly episode to two episodes weekly. According to Mervat Kharyallah, President of the country's Office of Public Programming, the show was created in response to al-Sisi's 'repeated warnings' of Fourth Generation Warfare in an effort to combat the myriad 'rumors and lies' then besieging the beleaguered nation.[76]

Presented in the form of a pseudo-historical documentary, the show glossed everything from the wars of Roman conquest, to the formation of Israel and, of course, the 25 January uprising and subsequent election of Mohamed Morsi. Framed within the context of an elaborate Western assault on the morals and security of present day Egypt, the narrative drew on many of the same far-right-wing American- and Russian-generated fake news stories as those deployed in Amr Ammar's *al-Ihtilal al-Madani*. In one episode, for example, it presented President Obama's historical speech in Cairo in which he emphasised America's history of religious tolerance, including towards Islam, as an ominous sign of the pending revolution. Following quickly on the heels of Obama's reference to Senator Keith Ellison being sworn in with a copy of Thomas Jefferson's Quran, the segment flashes to the seal of the Brotherhood before spelling out Obama's plot to upset the government and to install Mohamed Morsi into power. Other nodes of machination include the US-Qatari relationship and the impact of the latter on the Arab uprisings by way of *Al-Jazeera*. 'Bernard Lewis's plot to fragment the Arab and Islamic world' and of course secret US funding for online activists like Wael Ghonim also feature prominently. Beyond the cacophony of media, *Hurub al-jil al-rabi'* served to crystalise the country's greater, post-revolutionary angst into a neatly sequenced and richly imagined narrative. Populated throughout with video archives and textual references to a host of supposed US source documents – including notably academic presentations on Fourth Generation Warfare at the US Army War College where al-Sisi likely obtained his fascination with the subject as a student from 2005 to 2006 – the film shadows the coming crackdown on civil society leaders, journalists and the Muslim Brotherhood as a kind of vehicle for the rhetorical 'protection' of truth. As Sissela Bok observed 'those who believe they are exploited hold that this fact by itself justifies dishonesty in rectifying the equilibrium'.[77] And, indeed, a certain will for *veritas* appeared to underscore much of the rhetoric of Fourth Generation Warfare flowing from the offices of Maspero and the Presidential Palace. Comparable to a work like *La hora de los hornos* (The Hour of the Furnaces, 1968), the famed, multi-part documentary which sought to frame the troubled rise of Argentine nationalism against the freeze of neo-colonialism, *Hurub al-jil al-rabi'* stretched the truth to confront a history of perceived deception. The degree to which it was successful in this regard remains undetermined.

Conclusion

Egypt's military establishment could only be forgiven for assessing the phenomenon of Tahrir as the expression of a warzone not yet wholly understood. The country was in the midst of a 'fifth world war', if one accepted, as Putin's 'vizier' Surkov claimed, that world powers had embarked on the 'first non-linear war of all against all'.[78] With the region spiralling into violence, the urgency with which civic leaders and public intellectuals sought explanation for the unfolding events amplified the power of narrative. Yet, the 'discoursive community' they assembled for doing so came at the cost of new, undissolvable lines of division.[79]

Ironically, given the author's eventual disenchantment with the victors of Tamarrod, Alaa al-Aswany's post-revolutionary tome appropriately titled *Nadi al-sayyarat* (The Automobile Club) remains perhaps the quintessential relic of the post-revolutionary mood. Set during the time of British colonialism, the novel tells of how the indigenous servants of the expat social club finally take over the establishment and make it their own. As with the Tamarrod campaign, the novel's principal force of antagonism consists of Western interests and 'the Muslim Brotherhood, who were well known for their opportunism'.[80]

As János László has observed the social psychological reasons for why societies turn to identity construction through history are not yet fully understood. Egyptian response to the country's first revolution in the age of information suggests that the experience of social upheaval – or at least the perception of disintegration – is a likely catalyst. But another, equally potent force multiplier in the construction of identity is narrative. There is little room for the ambiguities of human experience in the work of narrative. As Amartya Sen wrote in *Identity and Violence* (2006), the notion that a person's identity as 'a woman does not conflict with her being a vegetarian, which does not militate against her being a lawyer, which does not prevent her from being a lover of jazz, or a heterosexual, or a supporter of gay and lesbian rights', does not apply to the logic of narrative.[81] In the hyper-mediated sphere of post-revolutionary discourse, the imperative to choose sides becomes irrepressible.

5

Media Wars II: Tunisia

Introduction

Four months after Sofiane Chourabi's disappearance in Libya on 8 September 2015, Tunisian news began relaying reports that the thirty-three-year-old journalist and 'cyberdissident', along with his camera man, Nadhir Ktari, had been confirmed dead.[1] The satellite channel *Libya al-Hadath* featured an interview with a 'Libyan jihadist' who saw Chourabi and his companion being led into a forest outside the city of Derna and executed. The witness even named the killer in his testimony, a Chadian, 'Abou Abdallah Dhayahi', who he said 'blew himself up a short time later in a suicide attack'.[2] Advocates for the missing journalists, including Ktari's mother, challenged the authenticity of the interview. The director of the Libyan channel, Mahmoud Al-Hadi, responded in turn that the recording was authentic and that the witness was being held by General Khalifa Hiftar.[3] One year later, following a string of bilateral conversations and a fact-finding journey to Libya by Ktari's mother,[4] the story began to shift as reports surfaced that Chourabi and Ktari were still alive.[5]

Paralleling the fate of Ansar al-Sharia leader Abu Ayadh, Chourabi's disappearance in Libya became one of the stranger sagas in Tunisia's ever-increasing, post-revolutionary web of information jujitsu. A writer for one of the country's few independent newspapers *al-Tariq al-Jadid*, Chourabi was an outspoken critic of government censorship under Ben Ali. Having been active in blogging the 2008 Gafsa protests for *Global Voices* and *Menassat*, he joined a select group of young bloggers from the Middle East for a March 2010 event in Washington, DC held by executives from Google, the Washington Institute for Near East Policy and Freedom House. Topics at the event included 'digital media's power in social movements' and 'political parties and elections 2.0'.[6] In April 2011, after a year of revolutionary turmoil, he returned to Washington to participate in the 19 April event at Freedom House, 'From Freedom to Democracy: The Next Generation of Democracy Builders'.[7] The announcement of a coming vote for the election of a Constituent Assembly the following October prompted Chourabi to focus his talk on the threat of an Islamist majority win. He emphasised the need for an expansion of the 'culture of democracy' in Tunisia and pressed the importance of centralising issues of gender equality and public liberties.[8] The 2011 visit corresponded with the launch of *Tunisia Live*, a site promising to become 'Tunisia's first English language website'. The announcement, however, was overshadowed by news of his ostensible connection to Freedom House – an association that, as Sami Ben Gharbia prophesised in 2010, was tantamount to the 'kiss of death'.[9] Within weeks of his trip to Washington even fellow bloggers were lamenting the notion that 'Sofiene Chourabi was financed by the USA'.[10]

For his part, Chourabi seemed to relish the controversy. He emerged as a regular commentator on the country's myriad news talk shows, becoming an outspoken critic of the Islamist-led troika government. In 2012, his profile was further amplified following an arrest for 'offending public morals'.[11] He and a group of friends, including women, were caught drinking beer on the beach. The incident converged with a politically charged report he filed for *Ettounsiya TV*, in which he engaged with the issue of smuggling along the Algerian border (a focal point in the emergent narrative of Ennahdha complicity with extra-territorial actors). As Eric Goldstein and Oumayma Ben Abdallah wrote for *OpenDemocracy* on the occasion of World Press Freedom Day, in May 2015, the report drew the ire of a federal judge who alleged, ironically, that Chourabi had been 'complicit' in 'disseminating false information likely to harm the public order'.[12]

Writing in relative obscurity for *al-Tariq al-Jadid*, Chourabi's foray onto the international media scene began amidst the opening salvo of unrest in Sidi Bouzid. As he described in a 2012 micro-documentary it was 'by chance' he found himself at the centre of protests on 17 December 2010. Having filmed the unrest, he returned to Tunis on 19 December and uploaded his material online.[13] His name and reputation, he said, spread rapidly at that point.[14] Western outlets picking up on the news in Tunisia began quoting him as a source, often in the same breath as their description of him as a journalist. 'Soufiane Chourabi, a journalist for the *Tarik Al Jadid* [sic] newspaper, was one of the first to begin documenting the uprising in Tunisia's interior', wrote Kristen Chick for *The Christian Science Monitor*.

> When Tunisians saw images of fellow citizens rebelling, they lost their fear, he says. He credits the videos of youths tearing down ubiquitous photos of Tunisia's autocrat as a psychological turning point. 'They needed someone to do that simple thing of taking down the picture of Ben Ali, and that was it. That released them,' Mr. Chourabi says. 'When Ben Ali's symbol fell, there was no fear.'[15]

Already part of a well-established byline, Chourabi's description of the 'no fear' moment crystalised what had been a distinct rhetorical motif in the cyberdissident movement for half a decade. From as early as 2005, the collapse and rise of symbols in the course of social transformation had been a major theme among dissident bloggers. The irony in Chourabi's remarks, however, was that by 2015 he too would emerge as a new kind of symbol. Vanished beneath the fog of war in neighbouring Libya, rumours about his ties to the US, his presence in Sidi Bouzid and his relationship with Islam converged into a new, counter-revolutionary myth, one that framed Tunisia's cyberdissidents as '*cyber-collabos*' (cyber-collaborators) in a greater US conspiracy to shape the Middle East. Flamed by members of the ousted Democratic Constitutional Rally party (RCD) and spread online through blogs, radio talk shows and Facebook,[16] le '*torchon électronique*' (electronic cloth) of the ancient regime, as one blogger described it,[17] threatened to engulf the cyberactivists in the same web of communications warfare that had helped spark their movement to begin with. As Chourabi's fate exemplified – and for which La Rue de Sofiene Chourabi and Nadir Ktari was later dedicated in Paris – controlling the narrative of one's identity and one's cause could become a matter of life and death in the no man's land of twenty-first century digital activism.

Inform and Resist

The history of the information struggle in Tunisia extends decades. It was Habib Bourguiba, leader of the country's independence movement, who, along with a clandestine cell of like-minded associates, founded *L'Action Tunisienne* (1932) an underground newspaper designed for the purpose of exposing French corruption and defending indigenous ideals in the face of an oppressive regime. After the future Minister of Foreign Affairs, Mohamed Masmoudi, was abruptly dismissed from the Political Bureau of Bourguiba's Destour Party for publishing articles critical of the postcolonial government in the party-aligned newspaper *L'Action* (1955–8), reform-minded editors of that paper abandoned publication and launched an entirely new venture – *Afrique Action* (1960). As Henry Clement Moore wrote, the initial charge of *Afrique Action*, soon to be renamed *Jeune Afrique* (1961), appeared to be editorialising 'about the vices of personal power'.[18] In 1972, with government censors trained on the myriad threats from the left, the nascent Islamic Tendency Movement (MTI), later to become Ennahdha, launched the journal *Al-Ma'rifa*. Salah al-Din al-Jourshi, a co-founder of the MTI and first appointed head of the National Constituent Assembly (ANC) in 2011, described the centrality of the publication to the ideology of the rising Islamist movement:

> While other parties were busy dealing with the effects of the crisis from the Sixties – the result of the failure of the country's experiment in socialism – a small group of individuals, in the early 70s, began meeting in one of the mosques of the capital forming the first nucleus of the Islamic movement. Its direction was formed by clear and simple orations of ideology. Even though some of the organizational elements of the Movement began migrating to the public realm [the public elite took little notice]. Lectures and gatherings were held in mosques and in the corners of mosques. The magazine *Al-Ma'rifa*, which forecast the appearance of the hijab before Tunisian women began wearing it in the mid-seventies, was published and sold at kiosks. Although suspicious, the powerful elite never saw it as anything more than a 'passing phenomenon'. They paid little attention to the seriousness of the issues we related or the 'significance' of the movement to the balance of power. It was only with the Iranian revolution and the overthrow of the Shah that the warning became real.[19]

The paper Chourabi worked for, *Al-Tariq al-jadid* (1980), as well as *Al-Rai* (1977), *Al-Mawqif* (1984) and *Muwatinun* (2007) were among the secular opposition newspapers that managed to publish under the Ben Ali regime despite, at times, significant efforts on the part of the government to block them.[20] *Al-Fajr* (1990), a quotidian iteration of *Al-Ma'rifa* and *Tunisnews* (2000) were among the Islamist-oriented publications that continued to publish despite intermittent censorship.[21]

The last and most recent stage in the country's decades' long information war emerged in the early 2000s, when, after becoming the first African nation to gain access to the Internet in 1991, the Tunisian government set itself on a collision course with online users seeking to expand their information horizons while gaining access to a new platform for self-expression.[22]

In 2005, amidst the furore incited by Ben Ali's successful bid to host the United Nations' second annual World Summit on the Information Society (WSIS), a group of dissident cyber-activists launched a coordinated campaign to bring attention to the hypocrisy of the government's claim to be a world leader in the expansion of Information and Communication Technology (ICT).[23] Paralleling the work of the 18 October Movement and the dramatic sit-in and hunger strike staged in the offices of the Order for Tunisian Lawyers by eight leading human rights activists and labour leaders,[24] the effort by Tunisian cyberdissidents to bring global attention to Ben Ali's repressive tactics flowed largely through the recently created website *Nawaat* (2004–), an independent online journal launched the previous year by a group of four Tunisians living in the diaspora.

Mohamed Zayani, who interviewed several founders of the group – which included Sami Ben Gharbia, Sofiane Guerfali, Riadh Guerfali (aka 'Astrubal') and Malek Khadraoui – observed that *Nawaat*'s sophisticated use of ICT quickly served to distinguish the site from the multitude of emerging online forums dedicated to Tunisia.[25] In advance of the WSIS, *Nawaat* helped coordinate an online campaign, '*Yezzi Fock* Ben Ali' ('enough is enough Ben Ali'), that featured users from around the world submitting images of themselves holding the slogan in front of the camera. The site, Yezzi.org, was cut off in Tunisia less than twenty-four hours after its launch.[26] *Nawaat* itself had been blocked in Tunisia since April 2004.[27] But using creative and viral imaging strategies the campaign managed to grab the attention of an international audience, with major news outlets, including *Al-Jazeera* and *CNN*, making reference to the campaign.[28] *Nawaat*'s later use of elaborate digital maps, including an interactive diagram of Tunisia's clandestine prison cell network; a geo-tagged 'digital sit-in' featuring video testimonies of political prisoners attached to the icon of the Presidential Palace on Google Maps; and a video mash up of the presidential airplane spotted by plane enthusiasts on runways around the world, showed evidence not simply of the group's political tenacity in exposing the corrupt practices of the Ben Ali regime,[29] but a willingness to forge through advanced computing and viral messaging strategies, a distinctive and compelling 'aesthetic of communication'.[30]

Still, the 'thingly element' of *Nawaat*'s creative campaign (in Heideggerian terms) threatened to limit the group's message to a horizon of their own design.[31] While shaped by activist-artists whose formation preceded the existence of the Internet, anxiety regarding the real world applications of the movement's virtual character became one of the most consistent features of the group's internal discourse.[32] In the lead up to the 2010/2011 revolution, regional observers attempting to describe the movement's political identity referred to a loosely sketched 'anti-colonial' narrative in which Internet censorship as enforced by Arab regimes reflects yet another stage in the long history of Western imperialism.[33] Leaders of the cyberdissident movement also advanced this narrative at times, tying their cause to the liberation of Palestine,[34] and, following the WikiLeaks release of stolen State Department cables, broader indignation towards the double standard of US calls for human rights protections despite evident violations in Guantánamo.[35] But the more constant node in the discourse of *Nawaat* activists was the singular call for the freedom of expression. As Sami Ben Gharbia said in his testimonial from March 2017 for the post-revolutionary Truth and Dignity Commission, 'blocking and censorship' (*al-hajb wa al-raq'aba*) and holding those responsible for who was behind it constituted the primordial charge of the cyberdissident movement. Yet, as he testified, 'to this day', the principals in charge of censorship activities in Tunisia remain unknown.[36]

As a country facing immense structural and economic challenges, unfettered online access may have seemed like a cause of minor import. But because the government's censorship programme was heavily classified, dissidents' efforts to expose it were seen as a direct assault on the very lifeblood of the regime. And indeed, the political weight of the anti-censorship campaign could be judged by the scope of the government's response.

As the greater *Nawaat* collective reassembled in 2010 to stage calls for 'manif22mai', a day of action planned to coincide with an international day of protest against Internet censorship on 22 May 2010, the group's social media campaign (which included, famously, the moniker 'Ammar 404', a reference to the error page that would appear in lieu of blocked websites and the mythical name associated with the 'censor-in-chief')[37] generated enough bluster to provoke authorities to pre-emptively fortify the planned protest site (the Ministry of Communication and Technology), while also arresting two activist (Slim Amamou and Yassin Ayari) who had advertised the event on their Twitter accounts.[38]

Few of the *manif22mai* participants were surprised by the crackdown. The next time around they held a flash mob in which participants were to appear at a chosen destination dressed in white. Authorities were alerted to the new plan and tried to stem the protest by arresting anyone they found wearing white.[39] But activists maintained real-time communications and successfully staged their flash mob at an alternative location. While the turnout was minor, activists roundly regarded it as a success not least because it captured international attention. *Al-Jazeera*, most notably, was among the major media outlets to report on the action.[40]

Cat and Mouse

Methods of censorship in the Arab world have long been multifaceted. As the editors of the technology page at *Le Monde* noted, the Tunisian practice of controlling online information was distinct from that of Egypt, for example. Whereas Tunisians had put in place an 'elaborate' system of network controls and password theft, the Egyptian government did very little to block sites, resorting instead to a more 'classical' method of directly detaining bloggers and activists.[41] The story of Khaled Said in this regard was emblematic of the potentially dangerous experience of Internet use in Egypt. Still, physical intervention was not unique to Egypt. Followers of the Tunisian cyberdissidance movement regularly point to the story of Zouhair Yahyaoui (aka 'Ettounisi'), an activist and founder of the satirical TUNeZINE who was abducted from an Internet café by Tunis police before being arrested and charged for 'knowingly putting out false news giving the impression of a criminal attack on persons or property' (a violation of article 306b of the country's old Criminal Code).[42] Yahyaoui spent over two years in prison before dying of a heart-related condition soon after his release, on 13 March 2005. His case highlighted not only the Tunisian situation, but a 'worldwide trend of repressive governments cracking down on Internet journalists and dissidents' in the early years of the twenty-first century.[43]

The increase in government censorship between 2005 and 2010 emerged concomitantly with the evolution of the cyberdissident movement. As the blogger Tarek Kahlaoui wrote for a critical piece in Nouri Gana's collection *The Making of the Tunisian Revolution* (2013) the online movement since 2007 had been moving ever closer to materialising its activity in the streets; a shift in the centre of gravity from expatriates living abroad to 'mainstream blogs' operating inside the country. Efforts to convey the events in Gafsa in 2008 through alternative and new media websites – despite the government's media blackout – were indication of the movement's potential.[44]

Mirroring the concretisation of the cyberdissident movement, Tunisia, along with other regimes in the region, began rapidly innovating its censorship capabilities. This included everything from Distributed Denial of Service (DDoS) attacks on cyberdissident websites, to the hacking of emails and account passwords, or the use of DPI, a technology that allows governments to 'track and filter everything passing through their networks'[45] Rebecca Mackinnon, a co-founder of *Global Voices*, described how DPI had become by 2008 a tool of choice for dictators around the globe. She notes that in 2011, after Egyptian authorities succeeded in shutting down the Internet for several days, a researcher for the US-based NGO Free Press, Timothy Karr, discovered that the Egyptian government had 'purchased DPI technology' from a California-based start-up, Narus, in 2005. The 'multi-million dollar deal', she notes, allowed Egypt's Giza Systems corporation to 'license' the Narus DPI technology to a 'variety of unsavory governments' throughout the Middle East and North Africa. Troublingly, the Narus program, 'NarusInsight', was awarded a US Homeland Security Award for Best Real-Time Dynamic Network Analysis and Forensics in 2010.[46] Sami Ben Gharbia highlighted the continued use of DPI technology in his testimonial before the Truth and Dignity Commission in 2017.[47]

The complexity of the Tunisian censorship apparatus – which included not just DPI technology, but email and Facebook password phishing, digital trolling, DDoS and, of course, physical incarceration – generated, in turn, a superabundance of creative alternatives for communicational resistance.

Sami Ben Gharbia highlighted the adaptive quality of the Tunisian cyberdissident movement in a speech he delivered for a 2012 conference in the Netherlands titled 'Curating Reality: New Tools for Investigative Journalism'. Pointing to the example of the stolen State Department cables which WikiLeaks began disseminating to major Western news outlets in late November 2010, Ben Gharbia (who describes Julian Assange as a 'friend') detailed how his website *Nawaat* worked to translate the material into French and Arabic and to 'contextualize' the information such that Tunisian censors would be unable to target the material.[48] Ben Gharbia has remained coy about the specifics regarding this and other information operations. He noted, for example, that he had not received the leaks directly from WikiLeaks but rather through an intermediary source – 'a leak within a leak'.[49] Such guardedness is to be expected, of course. As late as May 2011, Ben Gharbia was reportedly detained and interrogated by Tunisian authorities regarding publication of the leaked Presidential Action Plan on the *Nawaat* website.[50]

Seeking to expose the government's elaborate system of online censorship, cyber-activism in Tunisia, a strictly anonymous enterprise, centred rather on identifying the actors and methods behind the regime's covert surveillance technology. As the editors at *Nawaat* described in reference to the State Department cables, the aim was to expose the image of Ben Ali and 'the perception of the practice of power in Tunisia' (*la perception de la pratique du pouvoir en Tunisie*).[51] Their charge was to generate a clear and decisive counternarrative to the deceptive façade of the Ben Ali regime. Less clear in actionable terms was who or what would help delineate a narrative identity of their own.

The Strategic Aesthetics of Online Activism

Paralleling the work of cyberdissident bloggers, online activists in the *Nawaat* orbit created in advance of *manif22mai* a veritable ecology of protest iconography. Aesthetic resistance had been central to the cyberdissident movement from the outset. In his 2003 ebook *Borj Erroumi XL*, Sami Ben Gharbia wrote: 'the only path to the heart of the youth is an aesthetic one' (*le chemin qui mène au cœur de la jeunesse ne peut être qu'un chemin esthétique*). 'Political jargon lacks beauty', he added. 'It is through art that we might compel the youth to love the cause of liberty and justice.'[52] 'Azyz Amami', a pseudonym for one of the co-creators of *Nhar 3la 3mmar* and its current administrator, echoed this philosophy in a talk he gave at the Rencontre d'Arles in July 2011. The intention behind the page's viral imagery, he noted, was to maximise the 'emotion' of the image. Administrators would crop the parameters of a crowd shot to amplify the appearance of a gathering en masse or, more simply, Photoshop the image to 'make something beautiful'. This latter technique was particularly effective, he noted, because Tunisians in real life 'had very little to behold that was beautiful'.[53]

The phenomenon of art in the service of protest traversed the entirety of the Arab uprisings. As Marwan Kraidy observed in his seminal study of 'creative insurgency' – graffiti, dance, song and writing served to express 'rebellion' as much as they shaped it.[54] But *Nawaat* and later *Sayeb Sale7* and finally *Nhar 3la 3mmar*, which remains a hub of revolutionary organising (as of 1 January 2018, the page boasted 700,000 followers), were forerunners in the utilisation of digital aesthetics, a medium and a concept, which, in the first decade of the twenty-first century, was still in its infancy.

The genealogy of the group's techniques was eclectic. Contributors to *Nhar 3la 3mmar* point to the work of filmmakers like Romain Gavras or the digital *mise-en-scène* of an *Al-Jazeera* montage.[55] Interconnected and hyper-linked with a consortium of correspondent Vimeo domains, Tumblr and Twitter accounts, Google Docs, and Wordpress, the page was asymmetric in form and content but punctuated throughout with biting political satire and emotive imagery designed to maximise the spirit of rebellion. It incorporated local artists like the musician Mounir Troudi who recorded a two-minute studio clip advertising *Sayeb Sala7*,[56] and broadcast information about meeting points and protest strategies.

In one of the page's most poignant posts – a 19 December 2010 image of a 'see no evil, hear no evil, speak no evil' painting by an unnamed artist – a single comment signals the tide of history: 'Revolution Time'. The silence that follows (there are no posts between 19 December and 24 December) was deafening.

In certain respects *Nhar 3la 3mmar* mirrored the kind of aesthetic massification endemic to other comparable revolutionary experiments in the Global South. Like Augusto Boal's 'participatory' Theater of the Oppressed,[57]

Chilean protest murals, or art of the *extramuros* labour movement in Oaxaca, Mexico,[58] *Nhar 3la 3mmar* sought to 'vulgarize' human rights issues by extracting information from the 'sedate reports of Human Rights Watch or Amnesty International' and making them accessible for the 'average Internet user'.[59] Didactic in form and content, the activist-artists of the *Nhar 3la 3mmar* page were in the revolutionary business of '*pulling works out of museums and galleries* and bringing them into desacralized spaces'.[60] But like a 'disenchanted museum' such desacralised spaces diminished the impact of tradition as much as they expanded the reach of their message.[61] Tied to the medium of the Internet, the basic reproducibility of the content simultaneously expanded and threatened the inherent value of its meaning.

For example, on 1 February 2011, the day of the million-man march in Egypt, *Nhar 3la 3mmar* posted an elaborately doctored image of three masked protesters each holding the Tunisian flag. The text accompanying the image is drafted as a 'call' (*da'wa*) to 'neighborhood committees', compelling the people to unite as 'one hand' (*eid wahda*) in the face of 'saboteurs' (*mukharribin*).

While rhetorically invoking the Egyptian situation, the image was

Figure 5.1   R. M'Tmet on Nhar 3la 3ammar, Facebook, 1 February 2011[62]

Figure 5.2   Sayab Sala7 artworks on Nhar 3la 3mmar, Facebook, 18 July 2010[63]

derivative of an earlier post crafted in advance of *manif22mai*. The addition of the three masked protesters foregrounding a scene of explosions, a crouching soldier and Black Hawk helicopter, represented an escalation of emotion on the part of the artist-editor identified as 'R.M'Timet'. But it also served to hollow out the meaning of the original creation. Ungrounded, anachronistic and ahistorical, the image, like many on the page, multiplies and disintegrates before reassembling again in a different context.

Soon after its original deployment, a pro-democracy blog focused on Libya featured a version of the image with a Libyan flag in the centre.[64] The Iraqi-based online newspaper *Al-Zaman* recycled the image again, in April 2011, this time swapping the Libyan flag for a Palestinian one. Four years later *The Express Tribune* in Pakistan featured that version for an article concerning the threat of protests in the Gulf.[65]

No longer an expression of the Tunisian or Egyptian revolution *per se*, the image became a metonym for a broader kind of civilisational clash. Anti-imperialist and dystopic: the fantasy of an urban population at war with an invading military force could be virtually anywhere. Paradoxically, the Facebook hoodie is the only non-generic identity marker.

Aesthetic products born online often assume secondary and tertiary lifecycles as they move along the viral spiral. The early months of the Arab uprisings epitomised the rapidity with which locally inspired products in the digital age can find narrative coherency far and wide. For Tunisian activists, the phenomenon was a double-edged sword. Azyz Amami noted how the infusion of professional photographers several weeks after the start of the uprising undermined the efficacy of local protest journalists, in part because amateur collectives like that of *Nhar 3la 3mmar* could not compete with the reach of professionals and in part because the aesthetic composition of the new imagery, touching down briefly in Tunisia before departing for more distant shores, had assembled its own language of the Arab uprisings, one that was as much if not more tied to a Pan-Arab or diasporic palate than a Tunisian one.

Sami Ben Gharbia: The Art of Collaboration

The notion that cyberdissidents could reverse the course of history through a 'shattering' of the 'deceptive image of the regime' may have seemed idealistic, but it was not unprecedented.[66] Dean Wright, the Global Editor of Ethics, Innovation and News Standards at *Reuters* described the historic weight of precisely this phenomenon on the occasion of the twentieth anniversary of the collapse of the Berlin Wall. Joining a slate of high-profile investors, politicians and leading industry experts, Wright concluded his statement at the 'Breaking Borders' conference with a nod to the future. Repeating the success of 1989 would entail, he said, both an exchange of traditional professional tools (e.g. the dissemination of the *Reuters Handbook of Journalism*) and engagement with 'the new media reality'.[67]

Of no small coincidence Wright singled out the work of Sami Ben Gharbia. A political dissident then based in the Netherlands, Ben Gharbia had been on the radar of net neutrality and Internet freedom advocates for over a decade. Still, on the surface, his presence at the Breaking Borders conference, which was designed to celebrate the collapse of the Berlin Wall and the power of communications technology in the service of free-market economies, seemed like an anomaly. Having fled Tunisia in 1998 following a lengthy journey to Iran, his work had long been trained on exposing the 'neoliberal system' of 'global injustice'. His impassioned 2003 e-novel *Borj Erroumi XL* (named for one of Ben Ali's infamous and clandestine prisons) combined the classical genre of the Arab *rihla* (or travel narrative) with an ideological exploration of revolutionary Islam. Indeed his request for asylum in the Netherlands was derivative of the Tunisian government's suspicion concerning his visit to Iran, not Internet advocacy.[68] 'The independence of the Islamic Republic from the policies of World Powers is real and beautiful', he wrote. 'It is a fallacy to neglect this fact in the image we are told to construct.'[69]

Ben Gharbia's staunch opposition to neo-liberal reform agendas remained a defining aspect of his public discourse. But Wright (unaware or uncon-cerned of his politics) identified Ben Gharbia and the 2004 Harvard start-up *Global Voices* as leading exponents in the company's push to expand its reach particularly in the Global South.[70] Thomson Reuters – the parent company of the news agency – had formed a partnership with *Global Voices* in 2007 and Ben Gharbia as the organisation's Advocacy Director was at the centre of efforts to acquire original content from far off sources.[71] Simultaneously, Ben Gharbia was uniquely positioned to empower new African and Tunisian writers, who, as Wright pointed out, would otherwise be obscured by the 'cacophony' of the Internet.[72] The partnership was part of a trend, which included *Les observateurs* network of *France 24* and the new media desk at *Al-Jazeera*.

Breaking Through

The *Reuters* strategy as outlined at the Breaking Borders event came to a head on 17 December 2010. Standing on the soon-to-be hallowed ground of

Sidi Bouzid, Soufiane Chourabi, 'the only journalist present after Bouazizi's immolation',[73] was among a group of activist-bloggers closely associated with both *Nawaat* and Ben Gharbia's network at *Global Voices*. As such it was of little coincidence that *Reuters* became the first English-language news outlet, on 19 December 2010, to report on the protests, including Bouazizi's self-immolation. The story, 'Witnesses report rioting in Tunisian town', written anonymously (per *Reuters'* practice) and quoting unnamed sources, corresponded with a subsequent story on the *Global Voices* webpage by Lina Ben Mhenni, an early contributor to *Global Voices* and an associate of Chourabi. It was Ben Mhenni's post that provided the first full blueprint of the now famous narrative of the fruit vendor's death.

> An unemployed Tunisian set himself on fire in protest against his jobless-ness, sparking a wave of riots on the ground and solidarity and support on social networking platforms. While the fate of Mohamed Bouazizi, aged 26, from Sidi Bouzid, in southern Tunisia, remains unclear, Tunisian netizens seized the incident to complain about the lack of jobs, corruption and deteriorating human rights conditions in their country. From Facebook to Twitter to blogs, Internet users expressed their solidarity with Mohamed, who had graduated with Mahdia University a few years ago, but could not find a job. Being the only breadwinner in his family, he decided to earn a living and with his family's help, he started selling fruit and vegetable from a street stall. His venture gave him very little, enough to guarantee the dignity of his family. But city hall officials were on the look out, and have seized his goods several times. He tried to explain to them that what he was doing was not his choice that he was just trying to survive. Each time, his goods were confiscated, he was also insulted and asked to leave the city hall premises. The last time this happened, Mohamed lost all hope in this life and decided to leave it forever. He poured gasoline on himself and set himself on fire.[74]

Neatly sequenced along a clear progression of intra-connected causal events the emerging narrative mirrored a 'mythic topoi of grinding poverty', in János Lázló terms: a time-honoured pattern wherein the protagonist's 'rise-fall', gives ground to 'glorification' and finally the imaginative grounds of 'heaven-hell'.[75] Such 'recurrent patterns' of 'subordinate' and 'superordinate relations' have been shown by literary scholars for decades to underwrite the generic fundamentals of tragedy and comedy. The former (as Northrup Frye observed) can be identified in 'spatial-emotional' terms by the protagonist's isolation from society. (In comedy, the protagonist ultimately assimilates.)[76] In the case of Bouazizi the latter two elements of this mythical triad were established from the outset. To convey the image of the fruit vendor as an isolated individual forced to go it alone, 'netizens' and journalists needed simply to ignore any collateral ties he may have had; syndical, fraternal, familial, or otherwise. By attaching to his spectacular death the afterlife of an uprising, his place as a martyr in the popular imagination was secured. The first dimension, however, of a rise and fall, appeared vacant in the actual story of the deeply impoverished Bouazizi. The false insertion of his having graduated from university became in this way critical to his imagined failure in finding a job. With the invention of this factoid the narrative of Bouazizi integrated all three nodes of mythical topoi. Expressed as a tearing apart from society – opposed to a drawing towards it – the story was easily understood as a 'myth of tragedy'.[77] As one speaker later described on *Al-Jazeera* in recollection of the uprising, Bouazizi supplied in his death 'a beautiful painting, one that nobody could decode, one whose symbolism no one could understand'.[78]

In a repeat of the government's information blockade surrounding the 2008 protests in Gafsa, Ben Ali's surveillance apparatus at first attempted to censure reports of the incident. Ben Mhenni has refered to an Arabic language Facebook page called 'Mr. President Tunisians are Setting Themselves on Fire' that gained 2,500 followers in 'less than 24 hours' but was censored just as quickly.[79] Another example included the political activist and journalist Zuheir Makhlouf who managed to relay to the German-based *Assabilonline* video footage from an anonymous source based in Sidi Bouzid shortly before being assaulted and detained by Tunis authorities. *Assabilonline*, which was blocked inside Tunisia, posted the grainy video on its Facebook page along with a brief story about Makhlouf's detention on 17 December, the same day as Bouazizi's immolation.[80]

Direct coordination between cyber-activists on the ground and multinational news organisations, including *Reuters*, *Al-Jazeera* and *France 24*, became critical to breaking through the regime's surveillance apparatus. The government could shut down Facebook pages and silence local reporting, but confronting global media giants was a different problem altogether. However, similar to the façade of objectivity surrounding the *Atlantic*'s reporting on the early tremors of protest in Cairo the concealment of the bloggers' network was vital to the success of the collaboration. Moreover, the communicative imperative of narrative efficacy overshadowed the values of objective journalism. *France 24*, for example, where Chourabi was also a member of that channel's *Observateur* network, began its initial line of reporting on the events in Sidi Bouzid with an Arabic-language translation of the 19 December *Reuters* article. The two versions were identical but for one notable difference. In its byline on Mehdi Horchani *France 24* notes that he was 'a relative' (*aqarib*) of Bouazizi, not merely an observer.[81] Neither the *Reuters* article nor the *France 24* reprint, however, mention that Horchani was also part of the bloggers' network and had written about 'impoverishment and land seizure' in the neighbouring province of Regueb for a blog known as the Tunisian Cyber Parliament just six months earlier.[82]

Lina Ben Mhenni: Towards a Dominant Frame

Mohamed Bouazizi was not the first popular martyr forged through the nexus of citizen journalism and big media partnerships. In 2009, for example, Neda Agha-Soltan, the 'Angel of Iran' and 'martyr' of the 2009 Green Movement protests in Tehran, entered the global mainstream through what first appeared to be screenshots of a video uploaded to YouTube and distributed through *Reuters* television.[83] However, Bouazizi's story illuminated more clearly the potential for compelling (if violent) narratives to cut through the 'cacophony' of the Internet.

Consensus regarding the facts surrounding his death was far from immediate. Slimane Rouissi also a member of the *Observateurs* network who had contributed a post on the protests that summer,[84] produced an account of Bouazizi's death that was similar to Ben Mhenni's. However, Rouissi's article, which was published two days prior to Ben Mhenni's, also differed in significant ways.

> [Bouazizi] dropped out of school at a very young age (before high school) to help support his family of eight. His uncle had bought a small farm in R'gueb, near Sidi Bouzid, and his whole family moved there to work in

the fields. But the farm was one of those shut down due to corrupt land appropriations in the region. So Mohamed was forced to return to Sidi Bouzid to try to earn a living selling fruit and vegetables in the street.[85]

In certain respects Rouissi's reporting could be seen as foreshadowing the soon-to-be dominant frame as employed by *Al-Jazeera* that Bouazizi was a 'martyr' (*shahid*) for the cause of the 'oppressed' (*maqhur*).[86] As Sam Cherribi has shown, the Qatari-based media giant dedicated no fewer than a hundred talk shows to the subject.[87] Initially, however, Rouissi's framing of the protests in Sidi Bouzid as the extension of a land seizure dispute between farmers and local authorities, served in part to push back against the government's initial efforts to minimalise the impact of Bouazizi's immolation as an 'isolated incident'. Rouissi's backstory, in essence a social history woven around the aesthetic tropes of rural displacement, urban reinvention, local grift and, ultimately, cathartic transformation, elevated the singularity of Bouazizi's act by contextualising what made it unique. The once obscure history of the labour struggle in Tunisia – from the large scale strike by miners in 1937 that accelerated the drive for independence and the formation of the Union Générale Tunisienne du Travail (UGTT), to the Gafsa riots of 2008 – became increasingly common in popular as well scholarly commentary on the Tunisian revolution. As much a catalyst for the revolution that lay ahead, Bouazizi became a martyr for a history he likely never knew.

But while Rouissi's story of rural displacement, labour strife and the individual's descent into the informal economy had all the trappings of a familiar national saga, it was Lina Ben Mhenni's nuanced variation, gleaned from unverified Facebook posts and disseminated at lightning speed through the 24-hour global news cycle, that would generate the dominating media narrative of Bouazizi's life and death.

The most notable distinction of the Ben Mhenni narrative was the apocryphal detail that Bouazizi was a graduate of Mahdia University. The indiscretion has become a staple in academic scholarship on the Bouazizi phenomenon. The American researcher Merlyna Lim, for example, cites the '*Evening News*' on *Al-Jazeera* ('Mubasheer Arabic Channel') as having first reported the factoid as part of its interview with Bouazizi's 'cousin' Ali Bouazizi on the night of 17 December 2010. The interview was subsequently posted to Facebook, Lim notes, from where the story of the fruit vendor went viral.[88] Ben Mhenni, a college graduate herself when the uprising began, described being particularly shaken by the detail of Bouazizi's educational status. As she wrote in a subsequent post ('Et Mohamed Bouazizi s'immola par le feu') and recalled a year later in an interview with Ahmed Mansour on *Al-Jazeera* it was not the first time a young man from the region had burned himself in protest.[89] Just a year earlier in Montasir, she recalled, Abd al-Salam Tremish set himself on fire after being arrested by municipal police. 'But the news at first was saying that Bouazizi had a university degree', she noted. 'This spoke to me tremendously.'[90]

At first a recipient of the information, Ben Mhenni's 'horizon of expectations' became a pointed driver in her emotional attachment to the events in Sidi Bouzid. Her formation as a writer and a conduit to the West, however, helped to situate her at the crossroads of history.

As Ben Mhenni points out, her blogging began as a Fulbright Fellow to the United States where she lived for two years in Boston from 2007 to 2009.[91] Her return to Tunisia, where she pursued postgraduate work and taught Linguistics at Tunis University,[92] paralleled in many ways the historical shift in online activism in Tunisia from the diaspora to the homeland. Prior to 2007, as Roman Lecomte and others have observed, digital activism in Tunisia was almost exclusively clandestine with the bulk of 'cyberdissidence' being generated by users abroad.[93]

Ben Mhenni, whose father, Sadok Ben Mhenni, had been imprisoned during the reign of Bourguiba, wrote with an eye to politics from the outset. A prolific contributor to *Global Voices* and later *Nawaat* and *The Observers* much of her early work echoed the concept, as expressed by Dean Wright of *Reuters*, that local partner sites like *Global Voices* act as 'curators', siphoning and translating critical material from the local blogosphere in an effort to elevate key issues above the perennial 'cacophony' of the Internet.[94]

Her first article with *Global Voices* reflected this paradigm closely. Concentrating on the subject of censorship, she describes the case of Zied El Heni, a journalist who filed suit against the Tunisian Internet Agency (a symbolic gesture that was ultimately thrown out) to protest the censorship of Facebook which the government had blocked in late August 2008.[95] (YouTube, Daily Motion and other video sharing sites had been blocked for over a year.)[96] Among the distinctive qualities of her writing (and indeed that of *Global Voices* in general) was the markedly 'then and there' or 'observer' quality of her narrative perspective.

'Tunisian bloggers are rallying for a National Day for Freedom of Blogging on November 4', she wrote on 10 October 2008.

> It all started when Tunisian internet surfers welcomed with happiness the repeal of a ban placed on video sharing sites YouTube and Dailymotion. Many Tunisian bloggers celebrated this repeal of the ban by posting videos of songs downloaded from those two video websites on their blogs. But their happiness was cut short as the repeal did not last more than 24 hours.[97]

A mosaic of excerpts and quotes from other blogs, Ben Mhenni's work created the appearance of an information deluge, a technique, which, along with the translation of material into English, helped distance the sources from their creators. Simultaneously however, her work with *Global Voices* evinced many of the idiosyncrasies that distinguished her personal blog. This included the heavy use of emotive qualifiers, ironical enunciations, and above all, story: 'it all started when' may just as well have read 'once upon a time'.

Conclusion

When Tunisians finally gained access to Ben Ali's Internet Agency Headquarters (housed in the 'basement of a villa in Tunis') they renamed the space '404 Lab', a direct nod to the victorious campaign of the cyber-activists.[98] It was a symbolic gesture but also a personal one. Virtually all of those associated with *Nawaat*, *Nhar 3la 3mmar* and *Global Voices* had had their accounts stolen or their social media sites blocked.[99]

This final act of exposure, however, was also a culmulative act for the 'cyber-collabos'. Revolution aside, the operational objectives of their campaign had been to disrupt the regime's grip on information and communication technology. And now they were poised to expose the very infrastructure used for doing so. Was there a politics to this movement? An endgame beyond the free access to information, the right to assemble, the right to dissent? The answer to these questions is perhaps still open. But as I discuss in the following chapter, the immediate future left little room to manoeuvre. Consumed by a wave of rumours and misinformation – a phenomenon epitomised by the story of Sofiane Chourabi – the cyberdissident movement was in many ways forced to the sidelines of history by the very forces they sought to vanquish. As Isaiah Berlin once wrote: 'Large revolutions, attempts to upheave existing society and alter the course of events, do, at times, produce a break and change things deeply, but seldom in the direction which their initiators anticipated or desired.'[100] The resurgence of the old guard and the triumph of old ideologies – political Islam in particular – would fit this description succinctly.

6

Philosophy and Revolution

Introduction

'The reciprocal appeasement of the social and the political is the business of the old', wrote Jacques Rancière, 'an old business which politics has perhaps always had as its paradoxical essence.'[1] The years following the collapse of the old regime in Tunisia resembled this paradox closely. The demonstrations made famous by the coordination of youth activists soon gave way to an older fight, with the heroes of political movements past resurfacing like ghosts. Journalists and politicians returned from years in exile to assume power. Daily talk shows and news broadcasts drew reference to Habib Bourguiba, the leader of the Neo-Destour party that drove the push for independence from the French. Even the Islamist party Ennahdha, once persecuted by Bourguiba's nationalist agenda, now populated their website with allusions to 1956.[2] Ettakatol, the secular coalition party, touted reference to the year of independence on the tourism page of their website. 'The Ulemas were at the origin of the Personal Stature Code adopted in 1956', the page declared.[3] The ostensible message being: the Islamists do not have a monopoly on religion in Tunisia, cherished tenants of secular identity are also compatible with Islam.

The past, in such 'idealised' forms, seemed to efface the variabilities of the present. As Nouri Gana observed, with no shortage of irony, the tendency 'reached its crowning moment with the election of an eighty-eight-year-old, self-proclaimed Bourguibist, Béji Caïd Essebsi in 2014'.[4]

For people like Yassine Ayari, a well-known figure in the cyberdissident movement, the usurpation of power by leaders of the distant past was a point of sheer hypocrisy. '*Mon malaise vient de ceux qui font la politique*' (my discontent is spawned by the politicians) he exclaimed. 'When you critique them and when they arrest you. No! Not Marzouki', he wrote of the country's septuagenarian interim President Moncef Marzouki. 'He is a militant of human rights. But if you pursue politics like him watch out!'[5]

No longer restrained by the rigid delineation of power and ideology under the Ben Ali regime, the reversal and redeployment of historical icons from the country's institutional past threw into stark relief a complex struggle for power; one that was driven as much by mythology as it was class divide or wealth inequality. The question of who could claim access to the revolutionary narrative of 'the people' and how those outside its trajectory were cast became a source of wide debate. For the intellectuals of the postcolonial years, 'the generation whose political conscience was formed amidst the final throws of colonization'[6] – it was a moment long awaited.

The following chapter explores the response of one such septuagenarian intellectual, the philosopher and 'anthropologist of Islam', Youssef Seddik. Seddik's narrative of the revolution, I suggest, was emblematic in the sense that he sought to mitigate an older, more entrenched vision of secular identity with the ostensibly popular surge of political Islam, a phenomenon that Seddik, a celebrated scholar of Islam and an early acquaintance of Rached Ghannouchi, took personally.

As it became evident in Seddik's 'Chronique' – a frequent column that ran in the French daily Le Temps from the summer of 2011 to the spring of 2013, the spoils of free debate following the uprising proved more elusive than the obstacles it engendered. Navigating the paradoxes of democratisation would become as much an exercise in restraint as it was an experiment in free will. Bridging the gap between the two would emerge as the philosopher's greatest contribution to the ultimate confluence of narratives that would enable in the end a historic reconciliation and the country's first democratic transfer of power.

Staging Ground

Like many intellectuals of his generation, Youssef Seddik was living abroad when the Jasmine Revolution started, his exile having begun nearly thirty years earlier. He left Tunisia amidst the tense final years of the Bourguiba regime, in 1984, the same year in which clashes between protestors and security forces – the infamous 'bread riots' – resulted in the death of over 100 people. Once in Paris, Seddik, who had been working as a foreign correspondent for a Tunisian daily, began teaching philosophy, first at Paris III and then École des Hautes Études en Sciences Sociales. As he recounted in a series of interviews with Gilles Vanderpooten in 2014, Seddik learned about the imminent departure of Ben Ali from a Lebanese student who had been following developments on Facebook. 'I took him for a crazy optimist', he recalls. 'Two days later I was in Tunis.'

Once thrust into the storm of events, the atmosphere in the capital reminded him of Paris in May of 1968. With Ben Ali having suddenly and mysteriously departed for Saudi Arabia, the city, he wrote, felt alive with freedom. There was a sense of '*flux et reflux*', he wrote, 'fear and hope, depression and pride'.[7] Writing through the medium of the newspaper, Seddik devoted his investigations to the human dramatics of democratisation. His chronicle echoed some of the famed critical commentaries from a century prior, the work of Mohammed al-Muwaylihi or Bayram al-Tunisi whose satirical exposés on late Ottoman society sought to animate the struggle for democracy through colourful, often romantic prose in the classical mode of the *maqama*. Much like these early predecessors, his reflections on the revolution often drifted into a self-contained world abounding in Hellenic metaphors, allegory and satire. Above all there are good guys and bad guys in Seddik's chronicles. And none have a clearer role than the Islamists. 'Ghannouchi and his friends', he wrote, 'claimed for themselves the sky, leaving others the land, and I, nothing more than the horizon.'[8]

Rached Ghannouchi, co-founder of Ennahdha, was not merely the principal subject of critique in Seddik's chronicles; he was the philosopher's foil. Seddik's Ghannouchi, in his mannerisms, his words and his actions suggested more a typography than a biography. A crucial distinction as it was through symbolisation that Ghannouchi, who, in fact, we learn very little about in Seddik's writing, becomes a point of illumination in the otherwise murky waters of the philosopher's sustained meditation on the normative social impact of the revolution. Additionally, it is through typification that Ghannouchi, the character, helps to stage Seddik's drama. This effect is evident in one of the foundational scenes of the chronicle: the dramatic arrival of Ghannouchi at the Tunis-Carthage airport on 14 January 2011. Huddling amidst a crowd of impassioned adherents, 'the super sage of Ennahdha'

Seddik wrote, 'was greeted with the hymn of the Prophet Muhammad's flight to Medina by the sisters Najjar. "Here comes among us a full moon . . ."'[9]

Framing Seddik's characterisation of the emerging Islamist leadership was the revolutionary city itself – its thoroughfares and institutions, rumours and routines. The effect of his arrival quickly 'metastasized', Seddik wrote of Ghannouchi, his followers 'gaining behaviors, dialects, dress and personal appearance' almost instantaneously. 'Pious formulas invade conversations . . . scarves, veils, niqab and black gloves spread along with their specialty shops and their "stylists", beards of any size, pustules encysted in the foreheads of the young and the old.' 'The city', he wrote, had undergone a 'metamorphosis', becoming, in record time, a veritable theatre for 'this sad comedy'.[10]

As with the early twentieth century practitioners of the neo-*maqama*, Tunis constituted both the critical backdrop to Seddik's narrative as well as its principal scene of conflict. But it was also a theme that had long been part of his writing. '*Al-medina*' (the city), he wrote in *Nous n'avons jamais lu le Coran* (2004), represents the space in which tribal identity is supplanted by the Islamic space of shared ideology, or religion (*al-din*). But it is also the space in which 'debt' (*al-dain*) – a word homophonous in the Arabic to 'religion' – subordinates the older concept of 'polis-politès' to that of 'civitas-civis'. In the latter, the citizen exists in a state of indebtedness. In the words of Emile Benvèniste: the citizen in this model is 'bound to the territory'.[11] But in Tunis, like all colonial metropoles, the arrangement had been distorted beyond repair. The indebtedness of the Tunisian was to Paris, not Tunis, and the very architecture of the capital, Seddik noted, had been imagined to reflect this disenfranchisement.

> The capital of Tunisia is a large city of two million residents for whom there are no large spaces to assemble. The five or six squares it contains are functional – serving public transport for example. Even those with grass are not accessible to pedestrians. The streets alone belong to the people. In contrast, Cairo, where revolution also played out, all the squares, including Midan Tahrir, are popular and welcoming to celebrations and rejoicing. This explains in part why their revolution was different.[12]

The '*hausmannisation*' of Tunis had been construed in part to constrain political demonstration. Gatherings by necessity were forced to become 'rectilinear'. 'You have no choice but to stand behind one another, in files', Seddik observed. For this reason officials feared a massacre of historic proportions on 14 January, when grand columns of protesters amassed on Avenue Bourguiba. It was not just those in front who would 'take fire', he observed, but the entire gathering was subject to being trapped beneath the imminent 'stampede'. In the end, of course, this fate was avoided and the effect of the long march was the opposite: 'When he [Ben Ali] saw 500,000 demonstrators advancing on him, he realized he was the one who was trapped.'[13]

The interchange of social for political power as mitigated by the city would begin to elaborate, for Seddik, a principal trope in an increasingly consistent narrative of the revolution. 'Very often throughout history, it is the unresolved conflict between faith and the law of the City that becomes the source of catastrophe', he wrote in a three-part column from spring of 2013: *Matins et crépuscules de Dame Démocratie* (The Dawn and Dusk of Lady Democracy). The miraculous events of 14 January signalled not an end or usurpation of law by faith, in his view, but, rather, an interchange of value whereby the latter permeates or, indeed, infests the architecture of the city's social infrastructure. The effect for Seddik represented a cultural sea-change, one that was embodied, above all, by the very persons of the Islamist leadership. In bitingly sardonic tones, he captured the effect often, as in his 11 August 2012 column: 'Le sang de Monsieur Abdelfattah Mourou'.

> The hotel staff as well as the guests who had already arrived questioned his whereabouts in hushed tones as they watched the entrance to the grand building which opened onto the hell of the city drowned in the flames of the desert sun. 'The Shaykh has not arrived yet?' Then, less than an hour from the call to prayer, on an evening that refused to allow the burning walls and roofs the breath of night, the man arrived, invading the vast area of the lobby like a giant levitating cherub, his impeccable *djebba* embroidered with arabesques and lavished with holy words. The salon of the hotel seemed to redouble in size.[14]

Seddik's simultaneous fascination and scorn for the leadership of Tunisia's oldest Islamist party epitomised the country's most pressing post-revolutionary challenge, namely the collocation of a liberated Islamist public identity with the established, if beleaguered, legacy of Tunisian secular authority. Yet distinct from the polemics of the country's nascent revolutionary class, Seddik, who himself had led the ticket of a failed list of 'non-politicians, doctors, young engineers, professors, and pundits' for the Constituent Assembly elections of 2011, remained outside the corridors of power.[15] His chronicles of the revolution in this way offered an invaluable vantage point onto the 'structure of feelings' of the revolution, in Raymond Williams terms: the emotional and intellectual scaffolding of events that were at once extemporary yet indispensable to the emerging edifice of history.[16]

The Philosopher's Foil

Seddik began chronicling post-revolution deliberations shortly after the announcement of Constituent Assembly elections on 3 March 2011. Held the following October, the elections succeeded in selecting a 217-member body charged with the drafting of a new constitution. Ghannouchi's Ennahdha party dominated the vote, winning 40 per cent of the seats. But falling short of an outright majority, the party was obliged to form a coalition – or troika – with two secular-left parties, the Congress for the Republic (or al-Mottamar) and Ettakatol. Hamadi Jebali, the Secretary-General of Ennahdha, was appointed as Prime Minister. Moncef Marzouki of al-Mottamar was appointed President and Mustafa Ben Jaffar of Ettakatol leader of the Constituent Assembly. Yet, as Seddik wrote in his chronicle from August 2012, the real Chief of State remained the leader of Ennahdha, Rached Ghannouchi: 'celui par qui tous les scandales arrivent ou n'arrivent pas, s'il le veut bien' (he to whom all scandals arrive, or not, should he so desire).[17]

For many Tunisians, especially those born after the group's prohibition in the late 1980s, Ennahdha had been a largely clandestine phenomenon. Its members were dispersed across the country and its leadership was in exile.[18] Seddik, however, had personal history with Ghannouchi. He was well aware of how far the Shaykh of Ennahdha had come and how far he was prepared to go. In the early 1970s, they worked alongside one another amid the 'vaste cours' of the College of Mansoura, near Kairouan. Both were in philosophy. Seddik, a recent graduate of the Sorbonne, taught general philosophy, while Ghannouchi, also returned from a period of study abroad in Syria, taught Islamic philosophy.

Both Seddik and Ghannouchi were born in the rural hinterlands of

Tunisia. Seddik in the city of Tozeur in the southwest Governorate of the same name and Ghannouchi in a small southeastern village of Gabès. At a young age they were each pulled into the excitement of the Arab nationalist movements. Seddik describes his father as a 'militant anti-colonialist' and he points to the date, 5 December 1952, the assassination of the Tunisian labour leader Farhat Hached by *La maine rouge*, as the moment in which he gained conscience that 'great people, relatives and neighbors, had engaged in the "final battle" against those foreign bodies that had occupied our neighborhoods and play areas with their soldiering and war games'.[19] Similarly, Ghannouchi gained access to the cause of Arab nationalism by way of his maternal uncle, a Nasserist who inspired in him from early on an intense passion for the struggle against Western imperialism.[20] Both attended the elite Sadiki college system, a Franco-Arab hybrid model of education designed to empower a sense of national consciousness. And following university, both departed for foreign shores.

But it was there, while abroad, that the formation of the young philosophers began to diverge. As Seddik recalls, his departure for Paris as a student, in 1966, corresponded with the arrival of Michel Foucault in Tunisia that same year. Seddik enrolled at the Sorbonne, where he pursued a doctorate in Philosophy. A year later the debacle of the Six Day War fundamentally altered his course of study. 'I left the oral defense of my Master's (DES) completely stunned by the humor of one of my old professors', he wrote. The Arab defeat, which signalled for Arab intellectuals of the time a historical threshold of little comparison, had been treated with an air of banality by some within the French Academy. Sitting at a café on the *place* Auguste-Comte, Seddik resolved to radically reorient his course of study and to commit himself to a 'philosophical appropriation of the Book' (the Quran), despite the invariable challenges implied in such a task.[21]

> How could a Philosopher-Apprentice admit to such impudence, situating the study of such a Book along the same lines of research and inquiry that provided modernity its very visibility and legitimacy? The Quran, at that time, had its place along the shelves of the Islamic history department, and often as well within departments of Comparative Religion, but never in Sociology or Anthropology, or the so-called fields of the Humanities.

[This, despite the fact, that] Lacan himself, in his famous seminar on the fundamental concepts of psychoanalysis, had planned to demonstrate an experiment he had witnessed among one of his Iranian patients who experienced a psychosomatic episode, a paralyzed arm, by an unconscious fear generated by a memory from the Quran.[22]

Jarred by the decimation of Egyptian and Arab forces at the hands of the nascent Jewish state, Seddik would reorient his course of study towards a philological, or as he described it, anthropological analysis of Islam. While contemporary points of comparison existed in other traditions – Emmanuel Levinas's *Lectures talmudics* or the work of Pierre Teilhard de Chardin in the Christian context – Seddik understood his project as being revolutionary, a move that had seldom been attempted by intellectuals in the modern Arab context.[23] He was not entirely alone, of course. Other major intellectuals of his generation like Hassan Hanafi (b. 1935), Mohammed Abed al-Jabri (d. 2010), Mohammed Arkoun (d. 2010), Mustafa Mahmoud (d. 2009) or Mohamed Talbi (d. 2017) were also exploring new ways to analyse the foundational texts of Islam from a non-sectarian, scientific vantage point. Significantly, Ghannouchi – in the tradition of people like Hassan al-Banna (d. 1949), Sayyid Qutb (d. 1966) and Hassan al-Turabi (d. 2016), or in the Maghrebi context Abd al-Hamid Ben Badis (d. 1940) and Malek Bennabi (d. 1973) – also redirected his energy at this time towards a reengagement – or renewal of *ijtihad*. While the two Tunisians' methodologies would converge, their conclusions could hardly be more distinct.

Ghannouchi, living in Damascus at the time, describes being equally shaken by the 1967 disaster. Having decided already to reorient his personal journey from a discovery of Arab nationalism to an embrace of Islamism,[24] the defeat of Nasser's Arab army inspired in him a new philosophical embrace of Islam. Unlike Seddik, however, Ghannouchi's project was expressly political.

Upon his return to Tunisia, he enrolled in the University of al-Zaytuna, one of the oldest centres of Islamic learning in Africa. His Master's thesis on the life and thought of Ibn Taymiyya, ultimately published in Saudi Arabia under the title *al-Qadr 'anda Ibn Taymiyya* (Destiny in the work of Ibn Taymiyya, 1989), was a shot across the bow of not only the religious authorities at al-Zaytuna, but Tunisian society as a whole.

'In the hands of Mohammed Abdel Wahhab', the eighteenth-century founder of the proto-Salafist movement known as Wahhabism, Ibn Taymiyya's radical call for 'jihad and *ijtihad* (*al-jihadiyya wa al-ijtihadiyya*) as an attack on the institution of the Caliphate (and all institutions of established *madhhab* or jurisprudence) sparked a revolution that touched virtually every corner of the Muslim world.

> Just as in the West in the age of Renaissance, the Muslim world was stirred by a great awakening. Muhammad bin Abd al-Wahhab's message of *jihad* and *ijtihad* inspired an unbroken movement – from India and Pakistan (Dehlawi and Mawdudi) to Egypt (al-Afghani, Abduh, Rashid Rida, and al-Banna), Sudan (the Mahdiya movement), North Africa (the Sannusi and Jama'iyat al-'Ulema' movements), and the movement of Khayr al-Din al-Tunisi, Abd al-Aziz al-Thalabi, Muhiyi al-Din al-Qalabi, and Allal al-Fassi – to push the *umma* towards *jihad* against its enemies, to abandon the guise of tradition (*taqlid*) and to unite its divisions around the mystical origins of Islam and Islamic thought.[25]

The notable exception, for Ghannouchi, was Tunisia. In Morocco, he exclaimed, Abd al-Wahhab's revival of Ibn Taymiyya 'was met with the warmest welcome and highest scholarly esteem, the result of close affinity between the Moroccan Nationalist Movement and Salafism'. In Tunisia, however, the Maliki school of law (i.e. *al-taqlid al-madhhabi*) and the 'Sufist education policy and Ashari belief system that bound the country's political and religious establishment',[26] resisted the message of *al-jihadiyya wa al-ijtihadiyya* thereby instituting a sharp, or irrational divide between the civic-religious sphere of the *umma* and the national-religious sphere of *taqlid*. It was this division that would set the course for an unbroken path of resistance.

In the early 1970s, along with Hmida Ennaifer and Abdelfattah Mourou, Ghannouchi formed an Islamist association known as the *al-Jama'at al-Islamiyya*, or the Islamic Group. As Ghannouchi later recalled, the group began by organising in the lycées and mosques and later infiltrated virtually all of the universities and institutions.[27] From the platform of their newly created journal *al-Ma'rifa*, Ghannouchi wrote shortly after the Iranian revolution of 1979,

Islam is not simply a spiritual call. It is a creed (*'aqida*) and devotional service (*'ibada*). It is a political system (*nizam siyasi*) and a complete society. There is no difference between the material (*al-maddi*) and the spiritual (*al-ruhi*). And that is why the Imam al-Khomeini is at once a spiritual leader for the Iranian nation as well as the leader of its great revolution.[28]

Forged against the background of the Iranian revolution (1979), the assassination of Anwar Sadat (1981), and the imposition of sharia under the Nimeiry regime in Sudan (1983), Ghannouchi's project of political Islam constituted one of the fullest ideological expressions of the movement's underlying philosophy. As he expressed as late as 2014,

Islam was born political. The Prophet of Islam, Mohamed (PBUH), was not just a man of religion, he was equally a man of the State. He established a State, at Medina, with a political charter, what we would call today a Constitution. And it was pluralist, having been signed by Muslims, Jews, and others. Upon the death of the Prophet (PBUP), his companions pursued his political message and his vision of a State for the application of Islam. This has persisted throughout the course of Islamic history. The Caliphs, who were political leaders, were also the Imams of the faithful in the mosques. They considered their responsibility to be equally religious as it was political. That said, having no form of clergy, opposition always remains possible. After the death of Mohamed (PBUH), no one could pretend to have a mandate from heaven. The caliphs were not representatives of God on earth. They were not considered sacred. One could challenge them and challenge their authority. Political and religious. Still to this day everyone can propose an exegesis of Islam.[29]

Echoing almost verbatim his work on Ibn Taymiyya nearly three decades prior, Ghannouchi, the newly elected leader of Tunisia's governing party, forecast his party's political objectives through a richly devised ideological critique of power shared across the Islamist landscape of the Muslim world. More than a political doctrine, Islamism for Ghannouchi provided a 'narrative identity' insofar as it served to synthesise a neatly delineated sequence of events that was broad enough to encompass the mythical origins of *al-jihadiyya wa al-ijtihadiyya*, the colonial struggle for independence, as well as the present day writing of a new constitution.[30] But it also created in this way an implacable opening for bitter division. To borrow from Amartya Sen, in the currency of identity one is 'situated' inside or outside the circle of affiliation. There is little room for in-betweens.

The Identity Politics of Islamic Pluralism

This also describes why for Seddik the subject of 'identity' was of 'paramount importance' to the drafting of a new constitution. As he described in his first column for the chronicles dated 29 July 2011: 'Those who consider identity a faux problem are sorely mistaken.'[31] With Ghannouchi's ideology in full view he argued that the question of identity needed to be addressed in the Preamble of the central document.[32] (And, indeed, this is precisely what occurred, with a nod to the secular gains of the 14 January revolution and its 'martyrs' followed by an exhortation of Tunisian heritage including its 'enlightened reformist movements that are based on the foundations of our Islamic-Arab identity'.)[33]

Pluralism as a rhetorical cause célèbre became central to public discourse on the writing of the constitution. Ghannouchi himself was keen to utilise the phrase in the context of the greater arc of his narrative identity; however, its meaning denoted a function far more specific than common interpretation of the word. For the Islamists, 'pluralism', as a moniker for collective free will, has long been celebrated, in part, as an expression of sharia – God's path for man. They point to the Ashari doctrine of '*kasb*' (or acquisition), elaborated most famously by al-Ghazali (d. 1111), that holds free will to be one of God's sacred endowments to man.[34] As Ghannouchi wrote in his most famous work *al-Hurriyat al-ʿamma fi al-dawla al-Islamiyya* (Public liberties in the Islamic State, 1992):

> Freedom is in Islam, in its culture and its civilized experience. As a funda-
> mental value, it serves as the basis for the valorization and condition of mar-
> tyrdom, Islamic beliefs, and civilized society. Before the believer confirms
> his acknowledgment of the existence of God and the truth of Muhammad's
> message, he affirms himself as a free, rational being. The 'I' in the moment
> of consciousness and freedom decides 'I testify (I) that there is no god but
> Allah and I bear witness (I) that Muhammad is the Messenger of God.'[35]

The ascent of each Muslim to practise *tawhid* is understood as a gift, even a responsibility, for Ghannouchi. And in this way Islam, 'having no form of clergy', is always and at once dependent on the individual to interpret and practise his faith.[36] The result is an infinite horizon of interpretations. Such 'pluralism', for Ghannouchi, supports the project of *ijtihad* as it dates back to the time of Ibn Taymiyya. It also allows for an embrace of democracy.[37] When asked the question: is this vision of Political Islam [of infinite exegeses] relevant to a democratic world? he responded: 'Islam, we might say, is democratic because it is explicable in different ways . . . You could have one Islamic party in power and another, equally Islamic, in the opposition.'[38]

Youssef Seddik: Towards the Concrete

Seddik too had long argued for the defence of pluralism, but from a very different perspective. His professor at the Sorbonne, Jean Wahl, had been one of the first French intellectuals to elaborate the concept of pluralism with his 1925 book *Les philosophies pluralistes de l'Angleterre et d'Amérique* (The pluralist philosophers of England and America). Notably, Wahl's subsequent major project was a translation and analysis of Plato's dialogue of Parmenides, published in 1926 as *Étude sur le Parménide de Platon* (A study of Plato's Parmenides).[39] In the early 1990s, Seddik also translated Plato's Parmenides. It remains today the most complete version of the dialogue available in Arabic.

Virtually all of Seddik's writing can be seen as derivative of his meditations on the paradox at the heart of the Parmenides dialogue, what Wahl described, in light of William James' work on the psychology of religion, as the problem of 'the One and the Multiple'.[40] For Seddik, who devoted much of his scholarly energy to uncovering the Hellenic components of the Quran and early Islamic thought, its applicability was directly relevant to the idea of *tawhid*. This connection is most evident in his 2004 magnum opus *Nous n'avons jamais lu le Coran* (We have never read the Quran). Devised amidst the ever widening gulf between the so-called secular and Islamist spheres of Arab society, the book emerged as a powerful if quiet rebuff of the sectarian violence then roiling the Muslim world – from Algeria to Somalia, Afghanistan and Iraq. But the project was also many decades in the making and indeed, as he later described in his intellectual memoir *Qui sont les barbares? Itinéraire d'un penseur d'islam* (2005), his aspiration was to redress not simply modern iterations of religious misappropriation, but 'all the institutions, of knowledge, of power, of those who guard the temples where the sacred sarcophagi are housed'.[41] Paralleling Ghannouchi, it was a call for freedom of thought and, above all, exegetic pluralism.

In Seddik's work, the Quran is a sprawling masterpiece, its genius spawning centuries of readers. Yet its concrete meaning and the revelations contained therein remain, by definition, unattainable for the mortal mind. Despite this, the text (of course in Arabic) engenders in man an infinite compulsion to read. The effect, as Seddik wrote, recalled for him a line from the work of the Argentine writer Jorge Luis Borges: '*El encuentro fue real, pero el otro conversó conmigo en un sueño y fue así que pudo olvidarme*' (the meeting was real, but it came to me in a dream which is how I was able to forget about it).[42] To direct one's reading towards the concreteness of the text was to set the soul on an infinite course of repulsion and attraction, certitude and doubt. But it was precisely this '*aventure de la dialectique*' that Islam, in contrast to the ecclesiastical tradition, sought to sustain. A deeply philosophical work, the Quran for Seddik commands of its readers an ever evolving horizon of compulsive self-inquiry. Yet relentlessly throughout history the powerful have sought to diminish the ambiguity of its teachings for the purpose of supporting their own earthly authority. It was this dynamic, above all, that provided the rhetorical gunpowder for Seddik's critique of Ghannouchi and Ennahdha. What distinguished him from other voices of Tunisia's secular echelons, however, was the underlying coherence of his work with the spirit of the Islamists' own philosophy.

While Seddik's vision of free-will, derivative in large part from the Mutazila school from which al-Ashari (d. 935) had originally split, rejects the doctrine of *kasb* in theological terms, the thrust of his philosophy could be seen as wholly compatible with that of his former colleague. (Seddik has often been called the 'Martin Luther of Islam'. The title is perhaps better suited to Ghannouchi.) But it is the paradox within Ghannouchi's formulation of pluralism that would become in various forms the principal source of Seddik's disdain. The individual – or 'I' – must be understood as indivisible, or whole. Paradoxically, however, to exist as such, the individual must also be understood as unique, or partial, and thereby part of the multitude in which he exists. Determined at once to recognise the totality of Islam as both profound divination and a system of social organisation in the profane realm of the living, the 'latter day inquisitors' ('*inquisiteurs des temps présents*') of political Islam appear woefully unaware of this contradiction.[43] It is a point of 'superb ignorance',[44] for Seddik, that, once again, stems from a misreading of the 'Book's' essential ambiguities, including the paradoxical nature of individual ascension and collective adhesion. 'There exists among certain "Islamists" the deeply seeded notion that the Quran, for all of eternity, determines the conduct of man', he wrote in 2014. 'The problem is that they are committing from the outset errors of interpretation.'

> The first error is in the meaning of the words. The word '*hadd*' [limit], for example, appears in all works of Islamic law or jurisprudence. Not only is it not present in the Quran (where only the plural is shown: '*hudud*'), but its meaning is misunderstood. Whereas '*hadd*' means a 'norm' or 'limit,' they [Islamists] attribute to it an entirely different significance, i.e. 'a sanction'! A norm has nothing to do with a sanction.[45]

From the use of the hijab to prohibitions on alcohol, Seddik's work represented a systematic attempt to redress many of the most pervasive forms of public Islamic conduct by way of re-complicating the act of Quranic interpretation. Like Levinas's interpretation of the Torah, this included reading for the Hellenic dimensions of the Quran. 'I believed I could show how this Greek horizon', he wrote, 'completely hidden (*mis en absence*) on the surface of the Quranic text, or deliberately obfuscated, provided a false sense to precise words, a mutated portrait (*portraits travesties*) of known figures, a deviated course to identified events in the history of the ancient world.'[46] Such an understanding of the text rejects outright the transposition of its learnings to the far removed world of the present. Its meaning is inviolable and timeless in the most literal sense that it resists all 'historical dimensions' and 'political visions'.[47] To drag the text into the machinations of modern man by way of sharia is an act of 'bad faith' for Seddik. The revolution, in contrast, was an act of 'good faith' par excellence – a fundamental reshaping of society by people, for people, and in the name of '*la Patrie*' (the homeland).

To be certain, there was no small amount of controversy to his philosophy.

In the early 2000s, for example, he was the target of a fatwa issued by the Organization of the Islamic Conference for a comic strip he created illustrating the Quran.[48] His partial translation of the Quran (*Coran, autre lecture, autre traduction*) for which the comic strip was created generated an equal amount of negative attention. But it was unlikely that the quiet bibliophile was prepared for the media spotlight that would follow him upon his return to Tunisia. As debate raged surrounding the creation of a new constitution, Seddik, as one of the most animate defenders of a strictly secular articulation of national identity from his perch at *Le Temps*, quickly became a regular commentator on virtually all of the country's major news broadcasts and talk shows and was often seated *face à face* with a Salafist Shaykh or party leader from Ennahdha. In 2012, a small skirmish broke out when he was confronted by Salafists before a talk he was to give at the mosque of al-Zaytouna.[49] In 2015, the government dismissed the President of the High Islamic Council, a body administered under the state, after he issued a letter to the head of Tunisian public radio denouncing recent comments by Seddik about the Quran and comparing the philosopher to Salman Rushdie, who's life still remains in peril after the fatwa issued by the Ayatollah Khomeiny in 1989 for his novel *Satanic Verses* (1988).[50]

At issue in Seddik's work was not simply his intellectualisation of the Quran but his indignant stance on the public display of religion. Amidst the vigorous public debate preceding the drafting of a new constitution this issue became primordial for Seddik. And it served to signify a more profound mode of hypocrisy embedded not simply in the project of Ennahdha, but the practice of Islam in Tunisian society writ large. 'The wait was not in vain', he wrote sarcastically in March 2012.

> The Shaykh Rached Ghannouchi, general guide of Ennahdha and quasi leader of the country, recently gave a conference addressing the very heart of the problem that has gripped our country for months. It rests on an old alternative. A dichotomy as fallacious as it is dangerous: God or politics – to sum it all up in two words![51]

In condoning the discourse on sharia, or *huddud* – '*le retour du religieux dans le politique jusqu'aux séquences les plus banales de la vie quotidienne*' (the religious inflection of politics in even the most mundane spheres of life) – Ghannouchi and Ennahdha hastened the very false dichotomy their discourse promised.[52] Such subtleties of influence assumed an air of lesser import, however, as debate surrounding the drafting of the constitution came to a head. In a September 2012 interview with *Al-Jazeera*'s Teymoor Nabili, Ghannouchi explained his vision of pluralism as a system in which magistrates, chosen by a local council (*shura*), would administer law based on independent judgement. Because interpretations of sharia should and will vary, deference defaults to the highest circles of authority. In Egypt, the Muslim Brotherhood under then-President Mohamed Morsi sought to legislate a similar model, identifying al-Azhar as the seat of supreme judicial authority. In Tunisia, Ghannouchi and Ennahdha suggested this role would be played by al-Zaytouna.

But the institutionalisation of ideological principles proved treacherous. Ennahdha created a firestorm of backlash in June 2012 after proposing for the draft constitution an article prohibiting violations of the 'sacred' in public discourse or display. Coming on the immediate heels of an egregious attack on an art exhibition (*Le printemps d'arts*) by Salafist activists who accused the artists of displaying 'blasphemous' imagery, Article 3 was broadly perceived by secular critics as an attack on the freedom of expression. Ennahdha attempted to deflect criticism, but following the massive outburst of violence on 11 September 2012, surrounding the release of the shadowy online video *The Innocence of Muslims* (see Chapter 5) Ennahdha was forced to sideline its advocacy for Article 3. As members of the commission charged with producing the final constitution began debating the inclusion of a clause defending the ostensible antithesis of the idea, that is, 'the freedom of conscience', Ennahdha, to the surprise of many, offered little resistance. Supporters of the group may have at first feared such phrasing would create an opening for an assault on 'good virtues', as the representative Sobhi Atig said. As soon became clear, however, freedom of conscience, once understood in the Islamic sense as a mode of free will, fitted easily within the group's existent philosophy of *ijtihad*.

Conclusion

Hounded for decades by a repressive central government, the long narrative of the revolution for Ghannouchi and the Islamists was one in which the people struggled to bridge the divide between politics and religion. Like the Iranian revolution of 1979, an event Ghannouchi referred to often as a watershed moment for his movement, the aim was not merely 'to banish from the lands of our *umma* the bipolarity of Western liberalism and socialism',[53] but to break from the cultural conformities of 'Turkification'.[54] This would entail restoring civil society to the 'pre-colonial' model of a 'self-regulating community', in Nadia Marzouki's words.[55] A process, which, in Ghannouchi's words, would entail above all 'patience'.[56]

Few outside the Islamist crowd, like Moncef Marzouki who Seddik also knew from childhood as a classmate at Seddiki College, had the courage to engage the 'conceptual dichotomy' of Ennahdha.[57] Other more radical actors on the right challenged the group's legitimacy by their willingness to participate in government *ipso facto*. For example, the 2011 fatwa posted by Shaykh Abu-Muslim al-Jazairi (described in Chapter 4) denounced the 'increase in voices calling for the secularism of the state, and the introduction of a constitution' as moving away from 'the historically and physically proved Islamic identity of the country'.[58] For Seddik, however, the effect of the Islamists' embrace of the democratic experiment was more dangerous in subtle ways, particularly for those less familiar with the historical evolution of the Islamists' discourse. The '*rire*' or laughter of Islamists, as constituent members debated the validity of secular laws, infused the process with an air of banality, particularly when held against the gravity of sharia. The resultant 'depolitization of the youth', Seddik held, constituted one of the country's greatest challenges.[59] And, indeed, few of those who led the charge in the streets were willing to the suffer the 'drama' of governing in the post-revolutionary milieu.[60]

Beyond 14 January, the politicisation of religion, for Seddik, constituted the greatest obstacle to the creation of a national narrative of the revolution. There was no small amount of hyperbole to Seddik's position. ('If the Muslim world, from Jakarta to Mauritania, resolves one day to address the problem of religion in public space, it will open the way for a planetary revolution' he once wrote.)[61] But it was also one grounded in a certain inevitability of discursive interchange. As János László observed, there exists 'surprisingly little' research into the formal correlation between 'national histories' and 'national identities'. One clear point of resolution, however, is the dialogical concurrence of 'representational aspects' – the cultural signifiers and historical icons that assemble within one narrative trajectory or another.[62] Clear synergy existed between the Islamists and their detractors like Seddik. The most evident point of interchange, perhaps, was the topic of Palestine. 'Who today could deny the Palestinian conflict is integral not just to our history but our worldview', Seddik wrote in defence of the draft constitution proposal to permanantly prohibit the normalisation of relations with Israel.[63] But located in the context of the revolution and its necessary redress of the existent social pact, this too was expressed in the artifical mode of narrative. Like all other points of gravity the subject of Israel became complimentary to or antagonistic of opposing processes of resolution. Speaking at the annual conference for the commemoration of the 1948 Palestinian *Nakba*, Ghannouchi exclaimed: 'every development witnessed by the nation in the direction of freedom, justice and democracy, is a step in the direction of the liberation of Palestine'. Or in yet more blunt terms: 'the Arab Spring was inspired by the Palestinian resistance'.[64] Concurrently, for Ghannouchi and the leadership of Ennahdha, 14 January was a revolution fought 'in the name of religion'.[65] Its origin was not simply the state's history of repression in the name of fighting terror, but the co-optation of religion by an ostensibly secular elite. (Under Ben Ali, like Bourguiba, he noted imams were appointed by the state, which administered their affairs and could even 'dictate to them what to say and what not to say during the Friday sermons'.)[66] Thus, in stark terms, the otherwise common position on sanctioning all relations with Israel in favour of the Palestinian cause became, in the course of the post-revolutionary scramble to narrate the history of the revolution, yet another marker of identity politics.

Unlike in Egypt, where the new nationalist narrative subsumed through violence and co-optation virtually all other narrative identities, post-revolutionary discourse in Tunisia allowed for greater conjecturing on the cause and consequence of the uprising. January 14 in this way was not a singular event. Rather, as I show in the following chapter, actors well beyond the shores of politics found space enough in the rhetorical realm of the uprising to imagine the experience as a religious awakening, born not amidst the class struggle of an impoverished Sidi Bouzid, but in the nearby holy city of Kairouan. More importantly, 14 January was not historical in the sense, as Marx wrote, that history repeats itself but most often in the form of farce.[67]

Neither the ghost of the country's independence leader Habib Bourguiba nor any other historical reference shadowed the glare of the revolution's vivid reality. As historians look back at the work of Youssef Seddik and many others writing amidst the transition, they are likely to find mostly doubt, much like the Cynic searching with his candle in daylight for a dint of truth amidst an ocean of myth.

7

Jihad and Revolution

Introduction

A month into the Tunisia uprising the level of violence between police
and protestors began to escalate. In the interior cities of Kasserine and
Tala, not far from where the demonstrations began in Sidi Bouzid, nearly
two dozen people were killed by police over the course of four days.[1] In
Kairouan, even greater numbers of dead surfaced. In response to the violence,
on 13 January 2011, Ben Ali went on national television to announce major
reforms and a new round of elections to be held in six months' time.[2] These
concessions, he said, were necessary to stem the impending 'bloodbath'.[3]

The following day, the shadowy media organisation known as The
Andalus Establishment for Media Production released its most extensive
commentary on the uprising to date. 'O my Muslim brothers in Tunisia,
today, twenty three years have passed since the criminal tyrant Ben Ali began
his reign', the statement read:

> During these years, he and his guards have remained imposing heavily on
> your chest, and oppressing you, and made you suffer unbearable torture.
> Your agony and tragedy became more severe after their ruthless corruption
> of your religion and life has reached unbearable limits. The latest uprising of
> Sidi Bouzid and the subsequent wide scale movement of the masses came as
> a loud scream of a tortured victim in the face of his tormentor. The scream
> broke the wall of silence, which had shrouded in Tunisia of the Kairouan
> for so many long decades [sic].[4]

As the primary organ of al-Qaeda in the Islamic Maghreb (AQIM), The
Andalus Establishment for Media Production had established itself as a

markedly steady voice in an increasingly volatile public media sphere. As the revolution moved into its second year, however, and with turnouts increasingly large at Ansar al-Sharia events in Gafsa, Gabés, Kairouan and elsewhere, the ideological reach of *Al-Andalus* and other such outlets appeared to be growing. By 2013, it was estimated nearly 3,000 Tunisians had travelled to Syria – heeding the call of the AQIM emir Abu Musab Abdel-Wadoud (aka Abd al-Malik Droukdel) – to participate in armed jihad against the regime of Bashar al-Assad.[5] Many remained to wage jihad within the country and others still would embark for the shores of Europe. By the spring of 2013 it appeared Tunisia too, like Yemen, Libya or Syria, was on the brink of civil war.

Although focused largely on events in Algeria – the group's historic bastion of aggression – AQIM and its discourse on the Arab uprisings threw into stark relief the ideological extremes of a yet undetermined struggle for the hearts and minds of a nascent revolutionary people. In providing an analysis of the group's communiques over the first three years of the uprising, this chapter seeks to identify the narrative extremes of the Arab uprisings, those calcified regions of ideological conviction that, through their radical orientation, help to define more clearly the ideological centre. As in the above quote, one of the group's most persistent methodologies of communication came in the form of narrative – a measured, at times relentless insistence on locating events within a timeframe of their own design. More clearly here than virtually any other sphere of communication, AQIM's discourse on the uprising illustrates the irrepressible tendency towards 'emplotment' and the subsequent force of 'narrative identity' in the conformation of kinetic phenomena to ideological conviction.[6]

'The Islamic Maghreb'

Tanzim al-qaʿida fi bilad al-Maghrib al-Islami (al-Qaida in the Lands of the Islamic Maghreb, or AQIM), known previously as *al-Jamaʿa al-Salafiyya lil-Daʿwa wal-Qital* or the Salafist Group for Preaching and Combat (French acronym GSPC), emerged in 1998 as an offshoot of the *al-Jamaʿa al-Islamiyya al-musallaha* or the Armed Islamic Group (French acronym GIA) which had been responsible for some of the worst atrocities of the Algerian civil war (1992–2002). AQIM identifies 1990 as the year its movement began.[7] Its spiritual origins, it could be said, extend further to the early years of Algerian independence (1962) when an intense internecine struggle for authority beset the leadership of the National Liberation Army (FLN). Following the usurpation of power by Houari Boumediene, the country's first Minister of Defense, Mustafa Bouyali, a former commander in the FLN, broke from the party, reportedly incensed by his exclusion from the revolutionary council, and began a virulent campaign against the emergent 'Islamic socialism' of the Boumediene regime.[8] Following years in prison, Bouyali was killed by state security forces in 1987. In his wake, his followers formed the GIA, which began operations against the state almost immediately following the 1991 coup that ousted from power an ascendant Islamist parliament.[9] Its reach and aspirations became transregional over the course of the civil war. In 1995, the group sent fighters into Tunisia where they killed half-a-dozen police officers.[10] As violence by the group intensified, its ideology coalesced around the causes of jihad and *takfir* (excommunication of other Muslims), setting in motion yet another round of indiscriminate mass violence, including beheadings.[11] Foreign nationals and Algerian journalists who wrote in French were among the GIA's favourite targets. By 1998, more than 150 journalists in Algeria had been killed as part of the group's campaign to purge the country of '*al-taghut*', or 'the false god' of power.[12] By 2007, the group's leaders, then under the banner of the GSPC, had forged ties to the al-Qaeda franchise in Iraq and discovered that the anti-US jihad in Iraq was a persuasive recruiting point, especially for those estranged by the internal Algerian violence of the last fifteen years.[13] Its media outreach increased dramatically at this time with the launching of *Al-Andalus* in 2009. Echoing its campaign against journalists in the 1990s, AQIM announced the creation of the outlet as an effort to counter what it described as a 'dirty propaganda war' levelled by a 'satanic media machine' that 'slanders the mujahidin' by selling 'lies and deception to the Muslim masses'.[14]

The 14 January uprising in Tunisia became an immediate point of vindication in AQIM's existent narrative of jihad. 'Tunisia of the Kairouan' set the stage for what the group's elder leader and *shura* council member Abu Obayda Yusuf al-Annabi described as a 'blessed awakening'.[15] The 2014 documentary *al-Jaza'ir wa al-nafaq al-muzlim* (Algeria and the dark tunnel) contextualised the uprisings within a long, often confused history of French colonial oppression, Zionist conspiracy, and above all domestic corruption.[16] Significantly, unlike the various youth movements or the nationalistic narratives of the secular and military elite, AQIM sought to explain the uprisings within a singular and transnational gloss.

> The battle that you are fighting today is not isolated from the general battle waged by the whole Islamic *umma* against foreign and local enemies. The battle, intended to uproot oppression, maintain justice, liberate the Muslim lands from the conquerors, dismiss their apostate quislings, and implement the Shari'a is one battle. Neither freedom nor justice will be achieved without an Islamic government, which cares for the people's rights, protects honors, spreads justice, and implements the *shura*. This government will only be existent by performing jihad against the Crusaders and the Jews, and by toppling their agents, the treacherous rulers like Ben Ali, Bouteflika, Muhammed VI, Al-Qadhafi [sic], and others.[17]

For decades AQIM ideologues had emphasised the connection between the region's 'agents' and their perceived rulers in the West. The outbreak of revolution enabled them to fortify this message as a populist cause. Doing so, however, required framing the protests as a regional movement with transnational aims. Other narratives of resistance were hardly immune to the same play of reason. The notion that the cause for rebellion was the oppressive tactics of the region's authoritarian rulers and, by extension, the neocolonial practice (read here as 'crusades') of Western and Israeli aggressors, flowed through the ideology of organisations as diverse as the UGTT and Ennahdha, the Muslim Brotherhood and the Wafd.[18] But AQIM subordinated the anti-imperialist element of protest to the narrative of jihad. The distinction appears semantic: was it the West, Israel, and despotic regimes that constituted the central axis of antagonism, or 'Jews', 'Crusaders', and 'apostates'? In strictly figurative terms it would seem irrelevant, but as a mode of emplotment the effect is severe as the former intuits actual variables of policy (i.e. should the postrevolutionary state sever ties with Israel); while the latter envisions only mythical solutions.

The crippling effect of AQIM ideologues on the Arab uprisings was seen most vividly in the case of Northern Mali where, in October 2011, a slow-burning, largely secular Azawad (or Tuareg) movement capitalised on the momentum of the Arab uprisings to accelerate their push for an independent and Tamasheq-speaking state in the region of Gao and Timbuktu.[19] As Arab and Salafi-jihadist groups across the region – most notably *Harakat al-tawhid wa al-jihad fi gharb Afriqiyya* (The Movement for Oneness and Jihad in West Africa, MUJAO) – gained prominence on the ground, the ideological and transnational apparatus of al-Qaeda soon drowned out the narrative of the Tuareg rebels. As the leader of the Azawad Liberation Movement (Harakat Tahrir Azawad), Bilal Agh Sharif, said in a 2017 interview with *Al-Quds al-'Arabi*, the jihadist groups 'ignore the population and the national movements on the ground'.[20] In late 2012, al-Qaeda-backed fighters effectively expelled the Tuareg rebels from the region. A year and half later, in July 2014, the leader of MUJAO, Hamadou Wald Khairy, declared the city of Gao an 'emirate' of the Islamic State in Iraq and Syria.[21] Abd al-Malek Droukdel has described such schisms as *fitna* (or division) wrought by the policies of the Islamic State group. But the general trend of transnational allegiance – to ISIS or al-Qaeda – characterises the essence of the greater Salafi-jihadist movement. In its disconnect from reality on the ground, the simultaneous resentment and practice of auto-subjugation to a distant foreign master mirrors the schizophrenic self-narrative of AQIM.

Like other al-Qaeda affiliates, AQIM has tended to predicate their authority on a remote and distant power. This is practised through both their historical self-narrative pieced together by loosely drawn mythical and historical references (be it in their *noms du guerre*, their organisational structure or discursive allusions) and their actual governing structure. In one remarkable exchange harvested from the cache of letters discovered at Timbuktu, a feuding Mokhtar Belmokhtar and Droukdel engage the subject directly. 'You delude yourself if you imagine that communication with the Central Command of Al-Qaeda will be faster and easier than exchanges with the Regional Command [AQIM] that is in your close vicinity', exclaims Droukdel.

> For your information, since the oath of allegiance and to this day, we have received from our leaders in Afghanistan only few letters signed by Sheikh Osama – may God rest his soul – and by Sheikh Ayman, may Allah protect him. At the time, we also received few letters of Sheikh Atiyatullah and

Sheikh Abu Yahya, may their souls rest in peace, despite our numerous letters asking them to help us effectively manage the issues of Jihad here.[22]

With leadership based on a separate continent and communications all but severed, by 2012 AQIM's strict governing hierarchy appeared predicated on little more (and no less) than spiritual devotion to the idea of a reigning core. The revelation, hastened by the insistence of Belmokthar to communicate directly with central leadership in al-Khorasan, spoke to not simply a crisis of political integration but a 'breakdown in the signifying chain' such that 'material signifiers' of the present became incompatible with the imagined ethos of the group's ideological structure.[23] The fallout between Belmokhtar and Droukdel, as captured in the Timbuktu cache, was sparked largely by the ambitions of the former to deescalate mounting conflict with Europe by halting the group's lucrative kidnapping operations. Despite the imminent challenge posed by continuing such operations, AQIM leadership blasted Belmokhtar, then head of a subdivision known as the 'masked brigade' for acting independently of the central command's 'plan': 'to pressure the Crusaders' coalition to ease the pain of our brothers in Afghanistan and get satisfaction for other demands such as the release of imprisoned brothers and a ransom equal to the value of hostages who were worth far more'.[24] Such disconnect, however, between the ostensible pragmatism of the brigade commander and a preordained schema of action as dictated by regional and central command, while hardly unique to AQIM, appeared all the more pronounced in light of the group's rhetorical imbrication by the Arab uprisings. Here was a radically shifting landscape of 'material signifiers', some of which cast into doubt core features of the group's narrative identity.[25] That democracy was a Western creation, that the Arabs longed for an Islamic State, or that true change would come only through violence, all seemed increasingly unjustifiable in the first few months of the uprisings.

Going into the experience of the Arab uprisings, the rhetorical outlook of AQIM hinged primarily on a perceived ideological coda of distinct cultural and historical references, including many that were ostensibly detached from the contingencies of the immediate geo-political scene. From the Crusades, to the Battle of Badr or the Umayyad Caliphate of medieval Spain, the use of historical reference points constituted, in Jameson's terms, a 'signifying chain' of the group's narrative identity.[26] Unencumbered by the auxiliaries of contingent experience, such mythical rhetoric provided an ideological stopgap to the ever-diminishing ordinance of the distant core leadership as well as a sense-making device for target audiences that require a common sphere of cultural affiliation. Jeffry R. Halverson has explored this dynamic of jihadist communications at length. In a 2017 article he drew a comparison between the discourse of AQIM and that of extremist actors in the American context, noting:

> Details or plot points require no reproduction because the intended recipients have an existing and longstanding cultural familiarity with the myth ... Only a small fragment of it (perhaps even a single word, phrase, or date) needs to be recounted as part of the ideological work. For example, in January 2013, when the American gun-control opponent Alex Jones warned a pro-gun control British television host on CNN that '1776 will commence again', he was invoking an American myth (i.e. the American revolution) that was readily understood by American CNN television viewers.[27]

Once underway, the uprisings appeared to thrust into chaos many of the natural coherencies between the 'signifying chain' of al-Qaeda's ideology and its 'existing and longstanding cultural familiarity'. Rather than recalibrate the charge of their lucrative organisation, however, the key spokespeople for AQIM, like Yusuf al-Annabi, turned to the power of narrative to reframe the course of events within a discursive structure conducive to their pre-existent worldview. 'By God, how can you explain what is happening now in Libya?' exclaimed al-Annabi in a lengthy interview released on 7 July 2011.[28]

> Is it not an armed, blessed, jihad waged by the Muslim Libyan people against the tyrant Al-Qadhafi [sic] and his security brigades? It is an armed jihad with which the entire Muslim people have sympathized, and which was endorsed by all the honorable and freeborn. It is being covered by the Al-Jazirah [sic] satellite channel live, and millions of people are following it. Similar to this is what is now happening to our brothers in Yemen and Syria, where the Muslim rebels, as a result of continuous oppression and killings, find themselves forced to take up arms in order to defend

themselves and their honor and to overthrow the regime. Even in Tunisia and Egypt, we should not forget the great number of dead, wounded, and captive among the Muslim ranks. We will not forget the battle of the horses and camels. We will not forget that the flight of Ben-Ali came as a result of Muslims breaking into the dens of the tyrants, and their besieging of their palaces. It is wrong, then, to generalize and say that the uprisings were peaceful. Reality proves this to be untrue.[29]

By framing the uprising as an armed insurrection, al-Annabi defended the place of jihad as a vanguard expression of social upheaval. Grounded in the experience of Libya, this narrative, from 2011, appeared evermore ominous as a new wave of violence was spreading across the region. Equally vexing to the AQIM ideologue was the ostensible role of Western support for the demonstrations. Viewed by many as a radical about face, US support for the Egyptian uprising appeared to compromise the otherwise neat division between the 'tyrants' and the 'people'. AQIM often qualified the former as 'puppets' of the West. Still the group's pre-existent ideological schema ultimately trumped any discursive contingencies for nuanced phenomena. This narrative imperative became markedly clear in al-Annabi's July interview.

> Regarding the situation in Libya, it has taken a dangerous turn with the entrance of the Crusader nations to the battlefield. We did not want our people in Libya to seek the support of the Crusaders in their battle to liberate themselves from the apostate tyrant. There are many doubts now, and a suspicious role being played by NATO in an attempt to impose certain orientations upon the Muslim Libyan people. There are indications that indicate despicable blackmailing by the Crusaders in return for their air support. Hence, we call upon our rebelling mujahidin brothers among the sons of Libya to be patient in their fight to remove the tyrant. We urge them to adhere to their arms, strengthen their ranks, and develop their own capabilities so that the Crusaders will not be able to impose any dictations or conditions upon the people of Libya, and so that they prevent them from interfering in their internal affairs. They must also preserve their origins and their Islam, and their uprising from deviating from its legitimate objectives.[30]

Recorded, evidently, prior to the 19 March NATO intervention in Libya, al-Annabi's discourse reveals a degree of profound antipathy towards the prospect of a shared common enemy with the West. While a more pragmatic interpretation might mitigate this coincidence with a restructuring of priorities, as would be the case for American lawmakers debating the role of al-Qaeda in the war in Syria, the 'narrative identity' of the group was profound such that it prohibited *de facto* any scenario positing Western involvement as other than criminal. In this way, al-Annabi's message appeared to converge with the narrative emanating from the media centres of the old regimes (see Chapters 3 and 4). Distinctly, however, AQIM sought to frame the 'Western Crusaders' plot' against the revolutions as an act of co-optation; an attempt to use 'policies, media, and military might . . . to contain these revolutions and to work toward directing them to serve their own interests'.[31] This was quite different from the myth of 'foreign fingers',[32] for example, which sought to poison the uprisings at their core. AQIM instead professed admiration for the uprisings as independent and authentic expressions of a popular desire for change. In this way, the group sought affinity with the 'youth' of the Arab Spring. A 13 January 2011 communication dispatched anonymously on *Al-Andalus* even went so far as to quote Abu Qasim al-Shabbi, the early twentieth-century Tunisian writer whose poetry became a battle cry among youth activists in Tunisia and beyond. Yet, here too the group subordinates the 'youth' of the revolution to the 'mujahidin all over the world', who, as al-Annabi exclaims, are 'the vanguard of this *umma* in its effort to be liberated from the shackles of tyranny and oppression'.[33] Indeed the group's messaging to the 'youth' of the uprisings is as much a warning as it is a vote of confidence. Even the line from al-Shabi appears selective in its ideological bent, citing not the famed opening verse from the 'Will to Life' (*Iradat al-hayya*: 'If one day the people wills to live / Then fate must obey / Darkness must dissipate / And the chains must give way') but a more obscure selection from the end of the first verse: 'He who doesn't like to climb mountains / Will forever live among the hollows / The blood of youth roars in my heart'.[34] As with the group's commentary on the role of Western intervention, its outreach to the youth of the revolution was predicated on a well-defined narrative identity that would serve to frame, rather than illuminate, the events of the Arab Spring.

At the forefront of AQIM's response to the Tunisian uprising was the

experience of the Algerian civil war. That history, which included the ascendance of a democratically elected Islamist party (*al-Jabha al-Islamiyya l-il-
inqadh* or the Islamic Salvation Front) followed swiftly by a military coup to
unseat it in 1991–2, loomed large over the tense early months of negotiations
surrounding the creation of a draft constitution in Tunisia when leaders of
the newly victorious Islamist party Ennahdha began referencing the fate of
their early counterparts in Algeria as a precautionary tale. In one particularly
contentious instance, Rached Ghannouchi was captured on camera telling
a Salafi activist to 'be patient'. 'We don't want to go down their path', he
said, referring to the overthrow of the Islamist government in 1991 and the
brutal civil war that ensued. 'Today we have more than a mosque, we have a
Ministry of Religious Affairs. We have more than a single store. We have a
State.'[35] Whereas Ghannouchi's reference served an analogous function, for
the ideologues of AQIM the Algerian civil war constituted an integral part of
their narrative identity. 'France and the infidel West prevented the ouster of
the tyrants in Algeria. They supported the military Coup carried out by their
agents among the commanders of the Algerian Army', al-Annabi remarked.
'It is not possible to separate the recent movement of the *umma* from the
movement of its faithful sons in the various fronts and fields over the past
decades.'[36]

By reconstituting the history of the Arab uprisings within a decades'
long lineage of conflict, al-Annabi sought to establish a degree of coherence
between the events on the ground and al-Qaeda's own self-narrative. It is the
long narrative of 'Algeria' that sets into focus the group's relentless accounting of its violent activities:

11 Jumada al-Thani 1433: In the municipality of Tifrit, Tizi Ouzou
Province, the mujahidin detonated two bombs; 8 Jumada al-Thani 1433:
The mujahidin courageously mounted an ambush against a judicial police
patrol in the MeklaÂ region of Tizi Ouzou Province; 8 Jumada al-Thani
1433: The mujahidin mounted an ambush on a group of army soldiers; 4
Jumada al-Thani 1433: In the village of Bala'idin, Akerrou Municipality
(Tizi Ouzou Province), a land mine planted by the mujahidin detonated
and seriously wounded one soldier; 1 Jumada al-Thani 1433: The mujahidin remotely detonated a bomb in the village of Bou Zahr, in the Sidi

Mustapha Municipality, against a special forces unit. 22 Safar 1433: The mujahidin detonated a high-yield explosive device against a military truck in the vicinity of al-Zahif [sic ad passim].[37]

The overwhelming majority of AQIM communications, as with the above, were devoted to the recording of its guerrilla activities. Seemingly arbitrary, this index of violence disseminated by *Al-Andalus* appeared to create in its totality the bloody canvas of a country at war. Yet the myriad attacks the group claimed and often broadcast through footage uploaded to YouTube indicated little about the cause or objective of such violence. For that, the recipient of this information is obligated to reference implicitly the narrative of the civil war. The 'horizon of expectations' (*Erwartungshorizont*) of the group's communications illustrates,[38] in this way, an explicit iteration of what Ricoeur would describe, in reference to Hans Robert Jauss's *Toward an Aesthetic of Reception* (1982), as the 'dialectic' of narrative identity.[39] That is, because the group's communications, dispatches, videos and recordings are intensely grounded in the micro-realm of an ongoing guerrilla campaign, the recipient must create, in response, an imagined horizon of coherence, one that is entirely bound to his or her own expectations. Digital reproduction not only extended the reach of such communications but amplified the phenomenon of narrative identity insofar it necessitated on the part of the recipient a presumed coda of meaning. An index of attacks may resolve immediate information gaps for one man while indicating virtually nothing for another. The group's principal communicators appear loosely aware of this dynamic and over the course of the Arab uprisings were able to stitch together an 'intelligible configuration' of their activities on the ground in Algeria with a wider set of allusions to happenings around the region.[40] 'Glad tidings are followed by even more glad tidings, inspired by the crowds of Islamic youths gathering in the fields and squares of liberation and revolution [Al-Tahrir Square]', began one dispatch captured online in late May 2012.[41]

Their [the youth] perseverance and sacrifices have helped to make these events possible, as have the actions of the brigades of mujahidin scattered throughout the Islamic world. Here we see the defeats of the worshippers of the cross and those who support them, including our countrymen who serve as their agents, pile up everywhere. From brave and perseverant

Chechnya, to proud and lofty Afghanistan, to the Caliphate in glorious Iraq, to noble and resistant Palestine; from Yemen, the land of wisdom and faith, to Somalia, the land of the first immigration, to the deserts of Africa, the lands of the conqueror garrisons, the banners of Islam have begun above the nascent emirates.[42]

Beyond propaganda, the degree to which the turmoil in the region served to recast the group's war in Algeria as a global struggle and, conversely, the extent to which the greater regional turmoil – including extreme sectarian violence – had begun to evoke a certain implied resolution, namely, the formation of a satellite system of emirates networked by a singular and static identity, has remained largely unexamined by Western analysts. In its very fragmentation, the Arab world in a state of upheaval obligated for virtually any commentator, *ipso facto*, an imagined concatenation, or at the least, a sense of coherence. AQIM was no different other than the fact that as an outlawed organisation whose communications were largely autonomous, ephemeral and asymmetrical, its existence was predicated already on the fantasy of coherence. Its aesthetic, like its communications recycled from a remote and limited ordinance, appeared acutely 'postmodern' in the sense that the organic mission it once espoused as a 'revivalist social movement' in the late 1970s and 1980s,[43] determined to salvage the identity of 'conquered' or 'threatened' Muslim lands – from the Philippines, to Kashmir, the central Soviet republics, Bosnia, Somalia, or indeed, Kabylie, Algeria – [44] had by 2011 become something more of a 'multitudinous photographic simulacrum', in Jameson's words, a glossy nostalgia for a once authentic 'collective project' grounded in the aspirations of a 'redemptive historiography'.[45] Postmodernism, of course, as Jameson understood, was characteristic of 'late' or 'multi-national capitalism', a moment wherein the commodification of once national forms, signalled (for Marxist theorists like Jameson) a certain twilight of the Westphalian project and the rise of an unheralded global Empire. Yet the Salafi-jihadist project – launched (in some respect) not by a disenfranchised proletariat, but upper and middle class engineers, doctors and businessmen – [46] was in many ways predicated on a similarly synthetic and transnational aesthetic of the glorious past. In part an extension, in part a refutation, and in part a reiteration of preceding (and 'polysemous')

Salafist waves,[47] from the eighteenth-century revolution of Muhammad Abd al-Wahhab or that of Hassan al-Banna in the early twentieth century, the basic unifying identity of twenty-first-century jihadism demanded at its core a certain aesthetic recuperation of the imagined past. Refracted through the techno-aesthetic lens of twenty-first-century digital communications, satellite broadcasts and social media messaging, the 'pastness' of the Salafi-jihadist project had become, by 2011, all the more glaring in its 'pseudohistorical depth', to again borrow from Jameson's terminology.[48] The degree to which AQIM and other such groups sought to mitigate the postmodernisms of their identity with the modernist, state-building energy of the uprisings is uncertain. Though as the siege of Northern Mali, like the occupation of coastal Libya or most infamously ancient Mesopotamia would suggest, these actions were conceived less as a gesture of reification (a mirroring of Tahrir with Timbuktu, for example, as an effort to unify the group's imagined ideology with a material present) than a violent reaction to a yet unreconciled schizophrenia, in Lacanian terms: the bewildered state of engulfment in 'pure material signifiers'.[49]

To counter such bewilderment AQIM employed a strictly 'there and then' narrative perspective. While many organisations and individuals inside the Arab Spring capitals focused on the process of reimaging the legal infrastructure, reforming civil society and reallocating power, al-Qaeda entered the post-revolutionary discursive sphere with a clear set of directions. 'We see that the focus of everyone in Tunisia should currently be on the spread of *da'wa*' read an anonymous dispatch from October 2012. 'We advise our brothers from among the youths of the awakening in Tunisia to stay away from issues of *takfir* of groups and their notables. They should leave matters of *takfir* to the *ulema* [sic] whose knowledge and actions are trusted.'[50] This impulse to contain the extemporaneities of a people in flux was manifest in the cache of letters seized following the group's expulsion from Timbuktu by French and Coalition forces in 2013.[51] As with the cautionary notes of guidance to the people of Tunisia, Droukdel had expressed concern over exercising too swiftly a heavy-handed brand of sharia in Mali as he felt it would repel potential followers and precipitate foreign intervention, as indeed it did.[52]

This compensatory narrative strategy was indicative of a pending identity crisis which had been simmering within the ranks of AQIM for virtually the entire duration of its existence. Begun as an outgrowth of the Algerian civil war and composed in part of military detractors, the group's foundational *raison d'être*, so to speak, was in part pragmatic: the outgrowth of a perceived and unjust Western-backed system of local oppression and a 'history of violence' on the continent.[53] At its origin, AQIM sought to position itself as an anti-colonial force with the bulk of its effort focused on the so-called 'near-enemy'. This posture easily aligned with the principal charge of al-Qaeda Central. As shown in a 2013 survey of fifty-five communications between 2007 and 2012, the group's predominant material signifier of hostile antagonism is the perceived collaboration between 'crusader' (129 mentions) and 'apostate' forces (273 mentions), be they local or foreign.[54] Reference to this rhetorical axis appeared 40 per cent more frequently in 2007 than the period between 2008 and 2011. The reasons for this are myriad. The above-mentioned study, for example, suggests the high density of references to crusaders and apostates in 2007 may have reflected an effort on the part of leadership in AQIM to more closely align its rhetoric with that of al-Qaeda Central. More interestingly, however, the narrative thrust of the reference remained static in the face of the shifting political landscape. As AQIM sought to mitigate the complexities of the uprisings, its spokespeople revitalised their anti-colonial message by amplifying the degenerative impact on the region of neoliberal policies. In this way, it could be said, by avoiding 'reconciliation' with their own largely unsuccessful history of attempting to unseat the military regime of Algeria through violence, the group's reliance on its narrative identity hardened in the face of the uprisings.[55] Its 2014 documentary *The Dark Tunnel* was particularly expressive in this regard. Interspersed with its peculiar visual iterations of broadcast media reports, raw footage uploads, and testimonials, the video – aesthetically mundane on the whole – takes pains to denounce the economic failures of the Bouteflika regime. This includes the simple display of unemployment statistics; but also the juxtaposition of the unemployment figures with images of urban poverty. The aesthetic dynamic is subtle but expressive in-and-of-itself. Converging with the narrative identity of any number of anti-globalisation movements, sectarian and otherwise, AQIM cultivated its image as an anti-urban, rural-based entity, devoted not simply to rejecting the precepts of Western governance, but the very idea of modern capital development. For example, in a 2007 communique Droukdel wrote:

> Big foreign companies took hold of our resources and our internal wealth, they controlled our domestic markets after they forced us to accept the principle of liberating the exterior commerce, the use of the privatization laws, transferred the ownership of many of our strategic economic companies to foreign countries.[56]

Privatisation, in the narrative of AQIM, is closely linked to an even more pervasive ideology that rejects liberalism as a form of social decadence. In one dispatch, Droukdel describes the cosmopolitan condition as an expression of 'mundane life', an attempt to connect the miseries of the poor to the political programmes of liberalisation.[57] In another, an anonymous speaker describes how 'heroic mujahidin . . . AQIM fighters . . . successfully stormed the French mining region of Arlit in Niger . . . One of the most significant sources of uranium in the world.' The speaker declares: 'the companies of the Crusaders, which are stealing our wealth and exploiting our sons, must know that they are legitimate targets for the mujahidin, and they must leave soon.'[58] In a dispatch from 2009, AQIM Shura council elder Abu-Hayyan Assim articulates the message more explicitly still:

> This is a reminder to Muslims, mujahidin, and their supporters that the tyrants have realized that they are partaking in a war with no end in sight . . . The abundant military provisions exported to them by nonbeliever and apostate countries have done nothing to quell the fire of jihad in the hearts of Muslims who are passionate about their religion. They try to employ oppression and terror, whose slogan was said by one of the biggest criminals in Algeria: 'We are prepared to sacrifice the lives of three million Muslims in exchange for an Algeria living in modernity.'[59]

The condemnation of those who would fight for 'modernity' is offset rhetorically by the 'lions of Islam on the frontlines'. 'Mujahidin', whose sphere of existence are 'mountains, valleys, and sands', are juxtaposed with an enemy who sits 'behind a desk', absorbed in the life of 'European and Gulf nightclubs'.[60] In a post from 2009, the Egyptian-born operational expert for al-Qaeda Abu Abdallah Ahmed exclaimed: 'young men . . . look forward to recovering [your] leading role in saving humanity from the corruption of the man-made methodologies and the greediness of the usurper capitalism which came to its end and led the world to the abyss [sic].'[61]

Like other al-Qaeda branches, the physical and emotional lore of rural geographies was central to the narrative identity AQIM projected. The staging device served to amplify their opposition to the policies of modern economic reform while also locating their cause within the imaginative domain of their chosen target population. Located in the Kabylie region of Berber North Africa, AQIM, in line with al-Qaeda's original emphasis on restoring 'occupied' lands to their Muslim character, sought to aestheticise the Auras Mountains as both a staging ground of rebellion and a land of deliverance.

In one impassioned dispatch, a speaker identified as 'Abdallah al-Jaza'iri' exclaims: 'The banner of Al-Qaeda was raised in the mountains of the free by the hands of Amazigh [the Berbers]. These mountains still shelter the sons of the tribes. These mountains raise the banner of monotheism and humiliate the banner of pagan polytheism.'[62] A major front in the civil war of the 1990s and early 2000s, the Auras become a metonym for the Amazigh themselves, who, in turn, are assimilated into an imaginative arena synonymous with the radical cultural project of al-Qaeda. Rhapsodies for pre-industrial society are commonplace in Middle Eastern cultural productions, but unlike the aesthetics of *turath* (heritage), for example, a celebration of the past, AQIM attempts to portray the mountains and deserts of the region as an actual sphere of action for personal deliverance. 'I do not know the unseen world', al-Jazairi writes.

> And I say: O students of the United States and Paris, your plot will be foiled, God willing. You will receive strikes on your heads that will make you forget the way to the land of the Amazigh tribes. You losers know the intuition of the residents there and their common sense in understanding what you are ignorant about.[63]

Not unlike Droukdel's later call for an autonomous and 'Islamic' Azawad territory in the Sahel, reference to the Amazigh of the Auras, while remote from the actual politics of Berber identity, served to allegorise a place and a state of being, both apart from the ennui and struggles of modern life. The Chaambi Mountains on the Libyan border would become the scene of such imagined deliverance for the post-revolutionary generation of Salafi-jihadists in Tunisia.

Conclusion

The neo-ruralist narrative identity of AQIM was not without circumstance. All former colonies – most with just half-a-century of independence – the countries of the Maghreb have continued to experience staggering rates of inequitable development. The concentration of wealth in urban capitals at the time of the Arab uprisings was particularly glaring. A recent report by the United Nations Human Settlements Programme (Habitat) estimated that most of Morocco's economic activities and wealth are 'concentrated in five urban agglomerations', with 80 per cent of that in Casablanca alone. In Tunisia, the statistics are comparable, with cities producing more than 80 per cent of the country's wealth. In Mauritania, as well, the UN estimates 80 per cent of construction employment, 60 per cent of manufacturing, and the majority of public sector jobs are located in Nouakchott. The massification of the urban experience has followed at an alarming rate. Sixty-nine per cent of Morocco's population will live in cities by 2030 according to UN Habitat. In Libya the estimate is 83 per cent. And in Algeria, with an estimated population of 36 million in 2010, 34 million are expected to live in cities by 2030.[64]

Against this background, the adoption of neoliberal reform packages by Maghreb governments in the past two decades has generated no small amount of controversy. The Ennahdha-led government and the case of post-revolutionary Tunisia proved no different in this regard. A year into the Jasmine Revolution, the government allowed Qatar Telecom to purchase some 380 million Eurodollars in Tunisian bonds including 75 per cent of Tunisie Télécom.[65] They purchased 15 per cent of the remaining state capital a year later.[66] As in much of Africa, billions continue to flow into the region from China and Russia as well as the West. Perhaps most ironically, some groups in the US, like the Inter-University Center for Terrorism Studies at the Potomac Institute, have continued to advocate for deregulation and privatisation initiatives as an imagined solution for stemming the influence of al-Qaeda in the Maghreb. In staggeringly brash terms, a 2013 report issued by this group suggested that 'increased regional economic integration . . . would make it [the Maghreb] an appealing market for Europe and the

United States.' Loosely based on largely discredited theories of economic disempowerment and radicalisation, few proposals could more explicitly reinforce the narrative paradigm of AQIM.[67]

Born amidst the outskirts of the metropoles in Tunisia's impoverished rural hinterland, the Arab uprisings appeared instantly conducive to AQIM's existent master narrative. For a time the organisation sought to retool its identity as a revolutionary movement, trained both on the material usurpation of existent authority and the abstract mission of jihad. Al-Annabi's allusions to the poetry of al-Shabbi, or later, Droukdel's invocation of the early twentieth-century Libyan freedom fighter Omar al-Mokhtar, became in this way subtle attempts to expand the organisation's horizon of solidarity amidst the 'historical events' of 2011.[68] As the uprisings became more complicated, however, AQIM's discourse on revolution receded and their eternal counter-narrative once again gained focus. With the utopic adventure of Timbuktu aborted, attacks on mining installations, hotels, shopping malls and other centres of capitalist activity multiplied: all vulnerable symbols in the organisation's strange show of aesthetic revisionism.

8

The Speculative Fiction of Now

Introduction

'**N**o sooner had our revolution – in its myriad aspirations and manifestations – begun to reach the hearts and minds of the people than
writers within the movement began to pose an urgent question', wrote Taha
Husayn in 1954: 'where is the literature of the revolution?' For the reigning
Dean of Arab letters it was a rhetorical question. 'Literary insight will not
suddenly fall from the sky', he exclaimed, 'nor erupt from the soil because a
revolution broke out on July 23, 1952.'[1] The very impermanence of revolutionary times, he held, reinforces the ahistoricity of literature. This, despite
the modern requisite of the craft, the novel, especially, to convey subjectivity
– the dynamism of particulars – as Husayn himself so forcefully realised in
his modern classic *al-Ayyam* (The Days, 1929–73).[2] The 'ambition of those
who would bring about the literature they had in mind, one that would recall
the exuberance inspired in them by the revolution and the profound emotion
they felt upon its success, with photography and data in hand', generated,
as Husayn observed, 'more a "revolutionary literature" than a literature of
revolution'.[3] The tendencies, he noted, were formulaic and tied to existing
forms of expression. 'Never once did it occur to them in their discussions that
their literature, just like those failed and hopeless efforts from the past, was
the result of yet another decree from on high to deepen our understanding of
society and to put things in their place.'[4]

Husayn's 1954 essay, which was republished in a special edition of the
renowned cultural journal *Al-Hilal* in 2015, struck an undeniably confron-
tational tone at the time of its publication. Responding, in part, to a much
celebrated discussion between himself, the psychologist Abbas Mahmud

al-Aqqad and the famed playwright Tawfiq al-Hakim in *Al-Hilal* two years prior, Husayn continued to scorn al-Aqqad's position that the revolution would inspire literary innovation and that those striving to articulate not simply the 'spirit of the revolution' in political terms, but, as he wrote, the 'hidden spirit that accompanies the revolution in the soul', would invariably achieve a kind of cultural revolution.[5] Al-Aqqad, in his defence, expressed a more nuanced perspective. As he wrote of the July revolution: 'this new uprising is itself the result of developments in the arts and culture . . . All great events are resultant of what preceded them, just as they will transform what follows.' The revolution was not the beginning of a cultural revolution but the fullest expression of a radical transformation already underway.[6]

Cinema, in the 1950s, seemed for many the most likely venue for capturing the lived excitement of the moment. Already part of a wave of success (Egyptian cinema production had spiked from just fifteen films in 1938 to seventy in 1952 making it the fourth largest film industry in the world) Egypt's Revolutionary Regime Council (RCC) led by Gamal Abdel Nasser moved to integrate the Golden Age of cinema into the core of its national reeducation campaign. In 1946, there existed only 285 'regular' theatres according to Yves Thoraval. Following the State's takeover of the industry in 1961, the number increased by a third.[7]

Understanding that nearly 80 per cent of the population was illiterate at the time of the revolution, the RCC quickly identified the industry of 'talkies' as a point of critical importance to communicating the message of the revolution. After the inauguration of Nasser as President in 1954, the regime launched an official inquiry into the construction of a Cinema Palace in Cairo designed to seat 10,000 people. In addition to the above mentioned initiative of creating 100 new theatres annually, each with seating for 300 to 400 visitors, the government sponsored 'cultural caravans' to deliver films to remote villages across the country. In Alexandria and Cairo cine-clubs were created where, for little or no money, audiences could view select works of interest from around the world. In 1957, the government established the Supreme Council for the Protection of Arts and Letters along with the Organization for the Consolidation of Cinema, to channel funds and resources toward domestic production.[8] All part of the process of what would become known as the 'Charter on National Action', which included everything from the construction of the Aswan High Dam, to the retention of 'financial institutions, public utilities, transport (excluding taxi-cabs), industrial concerns, insurance, department stores, large hotels, the media and the press, export-import trade and the marketing of major agricultural crops', the takeover of Egypt's private film industry represented a financial boon to the regime.[9] It also generated an unprecedented wave of success for Egyptian filmmakers. Films were screened at the Yugoslavia Film Festival, Cannes, Berlin, the Soviet Union Arab Film Week, and festivals in Uruguay and Argentina.

Concurrently, the boom in film production drew some of the country's most prolific artists to the silver screen, where the technology of mimesis and the ideology of crowds converged. The work of realist filmmakers and writers – from Niyazi Mustafa, Salah Abu Sayf and Tawfiq Salih to Naguib Mahfouz, Bayram al-Tunisi and Abd al-Rahman al-Sharqawi – figured neatly into the regime's narrative of the revolution insofar as the world they depicted – often gritty, crime filled and unflinchingly critical of corrupt societal practices – highlighted that which the British and the monarchy had left behind.

Virtually all of the films from this time conclude on a triumphal note with the arrival of law and order in the form of a sharply uniformed Anwar Wajdi, or another leading figure from the actors' guild at Studio Misr.[10] Playing opposite the law and order protagonist there emerged a new and powerful character on the Egyptian scene, the *futuwwa*, or neighbourhood strongman. Discussed briefly in Chapter 5, the *futuwwa* – in its fictional form – represented an alternative, strictly indigenous form of authority. Made famous by Naguib Mahfouz through literature prior to the revolution, the *futuwwa* provided, in narrative form, a symbolic vehicle for the revolution's nascent mythology. Self-made, self-determined and traditionalist in outlook, the quintessential *futuwwa* of the Mahfouzian corpus rises to power through a normative sphere of social comportment wholly alternative to the colonial ethos of European modernity. Drawing on memories of the urban merchants and neighbourhood protectors he knew from his childhood, Mahfouz recast the identity of the *futuwwa* through its original, medievalist gloss of Sufism, chivalry and fraternity to become a new complex figure for an independent Egyptian nation. Most often played by Farid Shawqi, one of the country's biggest stars, the figurative motif of the *futuwwa* became one of the country's most enduring cultural icons.

Thugs Rising

The year 2011 presented a similar tableau for the projection of a new cultural horizon. Yet, 'gripped with a sense of despair, even hopelessness', as Husayn wrote of the '52 generation, writers discovered that the demand for a 'literature of the revolution' would not, in-and-of-itself, make it so.[11] Academics and journalists sought to hasten the moment. A surge of testimonials, edited collections and recycled texts with newly minted titles appeared on the shelves within months of the historic eighteen days. Hani Mustafa has noted that some filmmakers too 'reshot' parts of works made before the uprising to 'connect with and use the event'.[12]

The quantity and quality of film production in the immediate wake of the 2011 uprising was anaemic, the result of decades of steady erosion under the Mubarak years and what Walter Armbrust described in the early 1990s as the 'growing irrelevance of the ideology and institutions associated with Egyptian modernity'.[13] In 1994, ownership of Studio Misr was transferred to the Supreme Council of Culture and the Ministry of Culture and soon after that to the Holding Company for Housing, Tourism and Cinema under the Ministry of Investment. In 2000, it was 'rented' to a private company, El-Exceer. And in 2010, with the Studio reporting some 40 million LE pounds a year in losses, Egypt's greatest engine of culture production was put up for auction.[14]

Concurrent with the state's privatisation schema digital technology was challenging the viability of cultural industries the world over with an unprecedented wave of piracy, unlicensed satellite broadcasts and intellectual property violations. Speaking with the newly created *Mada Masr* e-zine in 2015, Farouk Sabry, Vice President of the Egyptian Cinema Industry Chamber, noted that the destabilisation of the region had further exacerbated the industry's steady decline as the demand for any form of cultural production in war-torn Arab countries – from Iraq, to Syria, Libya and Yemen – was minimal at best.[15] Remaining audiences had shifted online and the supply chain was vast with dubbed and remixed films flowing into homes from Turkey, Mexico, Brazil and the US.

Still, artistic expression in the Arab world dies hard. Amidst the decline in Egypt's state-sponsored film industry there emerged a louder, more self-determined and commercially driven form of cinematic production. No longer tied to the engine of the state, union membership requirements or quality control, the centre of gravity for Egyptian cinema, by 25 January 2011, had moved across the river Nile to Giza, where a rag-tag outfit begun in an apartment above the Sobki butcher shop in Dokki was quietly becoming the largest film producer in the country.

The story of al-Sobki Films is one of the most unlikely success stories of the Arab Spring. Begun in the mid-eighties as an informal swap shop selling imported Bollywood and Hollywood VHS tapes above their father's butcher shop, Mohamed and Ahmed al-Sobki ('slaughtering the competition') had become by some estimates the largest film distributors in the Middle East by the end of the decade.[16] In 1992, the brothers broke into the production side with the low-grade crime thriller *Uyun al-Saqr* (Eyes of the Falcon) starring Nour al-Sherif. This was followed quickly thereafter with *Mr. Karate* (1993) and a string of low-brow cult classics that helped crystalise the Sobki style: a mishmash of musical comedy, romance and Bruce-Lee inspired action-adventure. With little critical success (Mohammed al-Sobki, in his defence, typically points to the 2009 film *al-Farah* as an example of quality filmmaking), Sobki Film continued pumping out new and recycled material nearly every year without stop. Among the studio's most famous works was the 2002 cult classic *al-Limbi*. A fast, funny film with a booming soundtrack, Mohammed al-Sobki claims to have produced the film for just 2.5 million Egyptian pounds, a tenth of what it brought in at the box office.[17] Based on a dim-witted character from the slums who endeavours to create a bicycle rental business for tourists, the film thrust into the spotlight the Sobki image and idea of *al-shabiya*, a folk infused, cultural vernacular ('a smoke, a glass, a dance, a song') born of the city's peripheries where the *tuk tuk* and donkey vie for transportation needs and corrupt local officers routinely make the rounds for bribes.[18] The film launched its star, Mohammed Saad, into the cultural mainstream. Mobile phone shops and coffeehouses around Cairo opened with the name 'al-Limbi',[19] scores of imitations flooded the film industry and al-Sobki Films increased its production three-fold, releasing three films per year, on average, between 2003 and 2008.

Characterised by crude, colloquial dialogues, piecemeal plots and prolific sexual innuendo the brothers al-Sobki have regularly drawn controversy for their films. In 2014, the studio gained renewed attention following an order from the office of then Prime Minister Ibrahim Mehleb (who was later dismissed on grounds of corruption) that the film *Halawat Ruh* starring the Lebanese actress Haifa Wehbe be pulled from the theatres for inappropriate sexual content (Wehbe plays a character that appears based on the star of Giuseppe Tornatore's 2001 film *Malèna* which features a seductive female lead). Despite Mohamed al-Sobki's expressed support for the Tamarrod movement, ostensible representatives from the group launched a campaign to boycott al-Sobki Films in 2015.[20] That move appeared to correspond with a misdemeanor injunction (later overturned) against the film *Regatta* (as well as its lead actress) for sexually inappropriate content.[21]

Underlying the controversy, and success, of al-Sobki Films in the post 2011 context was the studio's brash financial independence. As the rest of the country's film industry all but evaporated al-Sobki Films continued pumping out new works at a breakneck pace. One reason for this is that al-Sobki Studio is an entirely self-funded enterprise, unhinged from the restraints of foreign backers or subsidies. As Mohamed al-Sobki has put it: 'All the money is ours.'[22]

The self-financing of al-Sobki Films, however, does not substitute for the commercial weight of the studio's artistic design. As with the rise of Studio Misr in the wake of 1952, with a high powered roster of bigger than life actors, the secret of al-Sobki Films' success has been star power. And the biggest star in the Sobki universe is Mohamed Ramadan.

Ramadan was just twenty-three-years old when the 2011 uprising erupted. His only notable credit before his breakout role in the Sobki production *al-Almani* (The German, 2012) was a small if complex part in the 2009 film *Ehki ya Shahrazad* (Tell me a story Shahrazad) where he plays a shopkeeper who seduces three sisters before the eldest one cudgels him over the head with a shovel and burns him alive in an act of jealous rage. Within months following his Sobki debut, however, Ramadan had catapulted to the heights of the Egyptian film industry. The release of his 2013 film *Qalb al-asad* (Lion heart) grossed over 10 million Egyptian pounds during Eid holiday, the most in Egyptian history.[23] By 2015, he had over 5 million followers on Facebook. Perhaps most significantly, Ramadan's films appear to bridge the social divide. Writing for *Huffpost Arabic*, film critic Mohamed Husayn described a scene of populist rapture as passerbys crowd around the nearest

TV screens to watch the final episode of the mini-series featuring Ramadan, *Ibn Halal* (2014). 'The microbuses came to a stop for over an hour', writes Husayn, while drivers and passengers 'packed into the ramshackle coffee shops of the de facto bus depot between 6 October and Cairo'.[24]

As with Farid Shawqi's portrayal of the *futuwwa* in the wake of the 1952 revolution, Ramadan's career quickly became synonymous with a single aesthetic trope: the figure of the *baltagiya*, or 'thug'. Ramadan's *baltagiya* cycle – including *'Abduh mouta* (The slave of death, 2012), *Sa'a wa nos* (Hour and an half, 2012), *al-Almani*, *Qalb al-asad* (Heart of a lion, 2014) and *Ibn halal* (2014) – mirrored closely the storyline of the classical Mahfouzian *futuwwa* narrative. The orphaned child (kidnapped in the case of *Qalb al-asad*, a rural migrant in the case of *Ibn halal*), raised by a life of crime and thrust into a string of conflicts that generate, in the end, a kind of casual redemption (including winning the heart of the rival clan chief's sweetheart), can be traced back to the earliest iterations of the genre.[25] As with the *futuwwa* films of the 1950s, the final resolution is hastened only after the intervention of the otherwise extraneous police.

On the surface it was of little surprise that al-Sobki Films would gain a foothold in the *futuwwa* aesthetic. Prior to the uprising the brothers Ahmed and Mohamed al-Sobki had done much to reenergize the subject by remastering and redistributing several of the key classics of the genre including *al-Futuwwa* (Salah Abu Sayf, 1957), *Futuwwat al-Husayniya* (Niyazi Mustafa, 1954) and *Rayya wa Sakina* (Salah Abu Sayf, 1953). But the market success of al-Sobki's production surge (including what Ahmed al-Sobki claimed to be the largest single-day box office sales in Egyptian history with the opening of the 2011 comedy *Shari' al-haram*) following the uprising took many critics by surprise.[26] As Hani Mustafa observed writing for the English-language *Al-Ahram Weekly*, al-Sobki Films, with its focus on the *baltagiya*, played 'into the hands of a counterrevolutionary zeitgeist eager to reduce revolution to chaos and systematically connect the workings of protests with the workings of crime'.[27]

As discussed in Chapter 3, such anxiety became diffuse in public discourse following the events of 28 January 2011. In particular, the disbanding of the police – an event which, as Mustafa recalls, was still eluding accountability in 2012 – became a rhetorical flashpoint for conjuring the significance of the *baltagiya.* The discourse of (in)security persisted well beyond the eighteen days of course. Writing for *Al-Masri al-Youm*, in May 2011, Mohamed Elmeshed discussed how the unwritten code of policing Cairo's vast slums had broken down: 'police officers complain that many ex-convicts and thugs previously hired by allies of the formerly ruling National Democratic Party engage in these crimes, and they seek refuge in shantytowns and ghettos that police have limited access to.' He quotes Mohamed Naim al-Qatawi, an editor of the magazine *Al-Busla*: 'in the past, lower-ranking officers played an intermediary role between their superiors and those viewed as "half-citizens" (poorer Egyptians) by upper classes'. With the collapse of the regime, the lower-ranking, often younger officers charged with policing the city's slums abandoned their task. In turn, 'the narrow alleys and intricate communication systems criminals had in place to protect each other, compounded with police passivity . . . amplified the impenetrability of some of these areas'.[28] The unravelling of Egypt's notorious, but powerful policing apparatus continued to dominate headlines over the first two years. Reports of vigilante justice, including lynching, compounded the impression of lawlessness. In March 2013, in response to the theft of a motorised rickshaw and the subsequent kidnapping of a girl, residents hung two men, the suspected culprits, upside-down from the banisters of an open-air bus station and beat them to death. The incident was captured on film, generating widespread condemnation, but, as the writer of *Al-Ahram Weekly* Ahmed Morsy notes, there was no official follow up.[29] Violence against women also became a subject of international scrutiny. 'Under President Hosni Mubarak, the omnipresent police kept sexual assault out of the public squares and the public eye', wrote Mayy El Sheikh and David Kirkpatrick. 'But since Mr. Mubarak's exit in 2011, the withdrawal of the security forces has allowed sexual assault to explode into the open, terrorizing Egyptian women.' Responses by officials from the Muslim Brotherhood on the issue proved troublesome. El Sheikh and Kirkpatrick quote one lawmaker from the Muslim Brotherhood as remarking at a parliamentary meeting: 'How do they ask the Ministry of Interior to protect a woman when she stands among men?'[30]

Paradoxically, the *baltagiya* narrative of al-Sobki Films served to accentuate the perceived lawlessness of the post-revolutionary moment, while also romanticising the most iconic of counter-revolutionary tropes from the eighteen days of protest in Tahrir. In the pages of *Al-Shorouk* and throughout the liberal media, the *baltagiya* functioned along the lines of the storied *agent provocateur*, imploding the revolution from the inside out. Plain clothed and armed with 'white weapons', the *baltagiya* emerged as a mercurial, hybrid figure moving effortlessly amid the dregs of society and the proverbial *'feloul'*, a once common description for 'remnants' of the ancient regime.

Their place in liberal media was undeniably nefarious. But their discursive significance also served an ironical function. The *baltagiya*, a word originally used to describe tax collectors in Ottoman Egypt and meaning literally 'he who bears an axe',[31] was at once an enforcer of stability and an agent of chaos in Tahrir. He was both a public and private citizen, the anonymous amalgam of an extreme regime loyalist and transient riff-raff. In the context of the eighteen days, the *baltagiya* served to exacerbate the popular divide between a cloistered regime elite annexing its security through cash transactions and a populace exposed to the forces of chaos.

Al-Sobki Films of course was not the only creative entity dramatising the emotion of 25 January. Nor were they the only ones to integrate the symbolic power and complexity of the narratives. Capitalising on the market appeal of offshore festivals hungry to feature stories from the uprising, directors like Ibrahim al-Battout, Ahmed Abdalla al-Sayyid, Mohamed Diab and Yousry Nasrallah swiftly generated a wave of new works, 'photography and data in hand', aimed at capturing the images, if not the emotions of 25 January. Ahmed Abdalla's *Farsh wa Ghata* (Rags and tatters, 2013) was particularly successful in visualising what Rancière has described as 'novelistic realism', a 'fragmented or proximate mode of focalization, which imposes raw presence to the detriment of the rational sequences of the story'.[32] Abdalla's film never approaches the national scene, nor the national story of 25 January. Rather, Tahrir and the uprising are mere backdrops to the quiet upheaval of the film's voiceless protagonist. Echoing an earlier work in the canon of post-revolutionary Arab aesthetics, Naguib Mahfouz's *al-Liss wa-al-kilab* (The thief and the dogs, 1961), Abdalla's film, which follows a convict, newly escaped as part of the country's mass prison break on the night of 28 January 2011, evinces a 'new regime' of the arts insofar as it seeks 'separation from the present of non-art', restaging the past as a kind of ordering of the present.[33] But set to an intermittent score of Moulid music and Sufi prayer, the convict-protagonist at the centre of Abdalla's film evokes not the raw brutality of the *baltagiya*, but, rather, the moral ambiguity of the *futuwwa*.

Appearing here, as well as in several other major works from the time, most notably, Yousry Nasrallah's *Ba'd al-mawqi'a* (After the battle, 2012) and the 2015 mini-series *Harat al-Yahud* (The Jewish Quarter), it was the picaro-cum-strongman figure of *al-futuwwa* who reemerged as a harbinger of social indignation. However, as with the historical fine line that has demarcated empirical studies of the *futuwwa* and the *baltagiya*, we see in the cinematic aesthetics of 25 January a kind of radical crossover wherein the strongman legacy of Ibrahim (the *futuwwa*) – enshrined in Sufi lore and enmeshed in the folk traditions of old Cairo – becomes supplanted by the *baltagiya*, a starkly secular figure of earthly enforcement.

The 'slippage' between these two terms has occurred in degrees over the course of at least the last century.[34] Wilson Chacko Jacob has argued that the distinction between the two evolved rapidly following a popularised incident in 1936 wherein a famous singer was murdered by a local businessman and weight trainer in Alexandria who had made a name for himself as a *futuwwa* but would become seen henceforth as a mere *baltagi*. Elliot Colla has shown how the pulp fiction market of the 1920s created an opening wherein the historical identity of the *futuwwa* gained a degree of market currency as a narrative device, but evolved in turn to reflect the dramatic requisites of the craft such that the 'irrational acts of society and individuals can be understood through reason'.[35] The *futuwwa* in itself thrived as an amalgam of seeming contradiction, both a 'chevalier' of the downtrodden and a brutal enforcer of his own self-interests. The traditional relationship between the *futuwwa* and the *baltagiya* reflected a degree of ethical 'corruption' from spiritual ambiguity, in the case of the former, to pragmatic brutality, in the case of the latter.[36] But the *baltagiya* of the post 25 January era conveyed only one side of the two-sided *futuwwa*.

Belal Fadl (discussed in Chapter 4) was among those chronicling this paradoxical new figure on the Egyptian landscape:

> There it was, a black Mercedes racing backwards down the middle of the street right before Galaal Bridge and just past the Opera House. The driver was indifferent to the insults being hurled against him as he puffed on

his cigarette like a cowboy .. unloading a barrage of insults against the 'revolution' as is the way with our revolution.[37]

Such blatant defiance of social order, Fadl recalls, now intersected tragically with memory of the uprising: the bridge he had marched, the chants in the air, the blood, sweat and tears of a people united in defence of liberty. Such defiance was convenient for the usurpation of authority. In the absence of struggle, however, the attitude had collapsed onto itself, re-empowering the kind of extreme disregard for social wellbeing once endemic under the ancient regime. This was the 'seven star *baltagiya*', high-end thuggery that was in essence indistinguishable from the mythology of Gamal Mubarak, the project of succession, and the menace of two Egypts, infinitely at odds and living atop one another in a state of constant battle. But energised by the experience of revolution, this new kind of thuggery was no longer in need of the police. As Fadl recounts the scene diverted towards terror as the driver descended to frighten away the young police officer who had attempted to confront him. As an expression of sheer authority beyond the constraint of the law, the 'seven star thug' intersected with the wave of violence then gripping the capital – from the massacre in front of the Maspero offices (10 October 2011) to the events of Nile City. But it was part of a spectrum – the logical most extreme of an otherwise degenerate opportunist: the paid informant, the petty thief, the hash dealer and playboy marrying for a few spare pounds from a dowry.

Then, like now, this narrative has proved to be massively popular. As with the 1950s, however, its proximity to the experience of mass protest raises principal questions not simply about the enduring dialecticism of 'the people' versus the 'Popular' as Néstor Garcia Canclini articulated – but of the interdependence of narrative stasis and social dynamics – the implacability of the latter in the face of the former and the ever steepening prospect of true social change in an age of instant entertainment.[38] The commercial narratives of the *baltagiya* have stripped bare the notion that the popular elicits an emotion of 'constant obsolescence',[39] an ephemerality of experience tied exclusively to the market. Unhinged from the ambiguities of the *futuwwa*, the popular *baltagiya* motif represents, arguably, a pinnacle-like expression of the current wave of global populism: it is anti-woman, anti-intellectual and ruthlessly uncomplicatedly. The *baltagiya* motif is the perfect aesthetic accoutrement to the new strategic axis between Russia, Egypt and the Trump Administration. As with Trump's proto-populism, the ideology behind the *baltagiya* aesthetic appeared only remotely deliberate – emerging, rather, as the opposite: an anti-art, devoid of conception and driven instead, instantly and ubiquitously, by the singular weight of the dollar.

Speculative Realism

It is worth reiterating in line with this last notion that the typology of the *baltagiya* was born amidst the experience of a social transformation extending well beyond the shores of Egypt. Prior to 17 December 2010, the image and the connotation of the *baltagiya*, loosely writ, had been cascading through social media networks of the greater Mediterranean for years. Witness Romain Gavras, the Greek-born French artist whose powerful, incendiary music videos captured the imagination of the Tunisian cyberdissidents. Slim Amamou attempted to describe the emotion of discovering the filmmaker's work for the first time in 2008. *'Je fus impréssionné par ce petit chef d'oeuvre'* (I was impressed by this minor masterpiece) he wrote of the short film *Stress*. 'One thing in particular fascinated me: after viewing the clip, the message was clear . . . But I didn't know what it was.'[40]

Gavras had rocketed to stardom by the autumn of 2010. 'Stress', a song by the electronic music group Justice boasts half-a-million views on Vimeo. Its high energy, pseudo-documentary take on youth angst in the suburbs of Paris 'made the cover of the *Le Monde*' he recalled in a 2010 interview with *The Guardian*.[41] His most iconic work, the nine-minute music video to MIA's song 'Born Free' has been viewed 4.6 million times on Vimeo and nearly 4 million times on YouTube. The latter took the video down at first for its graphic depiction of police violence (a white-skinned redhead boy is pulled from a group of other white-skinned redhead migrants, ostensibly on the US-Mexico border, and executed at point-blank range by a border guard. It is a disturbing and blatant dénouement to an otherwise ambiguous storyline: imperialism murders even its own). 'No Church in the Wild', a 2012 video set to (and featuring) Kanye West's and Jay Z's powerful anthem by the same name, represented a full-circle for the artist depicting not simply the *fata-ayyar* of global *maqhur* – the disposed youth of global oppression

– but the aesthetic of revolution itself. Staged with barricades, Molotov cocktails, tear gas and the signature masked warriors of 'Stress', 'No Church in the Wild' synthesised a generational calling: from the Battle for Seattle, to the streets of Athens, Palestine, Egypt and beyond.

Ahmed Khaled Tawfiq and the Dystopic Turn

In literature, one of the Arab world's most popular, if understudied, novelists was also beginning to fashion a literary aesthetic comparable to the emotional anarchy of a Romain Gavras video. Primarily a young adult and fantasy writer prior to 2009, Ahmed Khaled Tawfiq began his foray into novel writing with a starkly cynical vision of a coming insurrection. Since the 2011 uprising, he has gravitated towards an ever more idiomatic aesthetic, turning to classical science fiction themes like time travel (*Ta'thir al-jarada*, *Mithl Ikarus*) and immortality (*al-Sinja*, *Fi mamar al-fi'ran*). But it was his 2009 blockbuster *Yutubiya* (Utopia) that remains the quintessential expression of the transcendental brutality that is the twenty-first century *baltagiya*.

The protagonist of *Utopia* – hair 'shaved on the sides, with a tall purple tuft in the middle, like a rebellious wild rooster' – strikes an image of creative delinquency in the tradition of Stanley Kubrick's infamous protagonist 'Alex' from *A Clockwork Orange*. Kubrick described the Hobbesian qualities of Anthony Burgess's dystopic novel: 'I don't think that man is what he is because of an imperfectly structured society, but rather that society is imperfectly structured because of the nature of man.'[42]

Tawfiq's nameless protagonist – like Burgess's and Kubrick's Alex – is an adherent of 'ultra-violence', a predator from the cloistered walls of 'Utopia' who goes hunting for body parts among the 'Others', inhabitants of an alternative society known simply as 'Cairo'. Yet, also, like Alex, Tawfiq's protagonist, speaking in the first person, aspires for brutal honesty. Tattooed, pierced and radically grounded in the immediacy of self-indulgence – sex, drugs and violence – the protagonist of *Utopia* is an uncensored insider of the powerful elite, but also, simultaneously, a vivid manifestation of the violent schizophrenia feeding the social divide. Tawfiq's protagonist exhibits the opposite extremes of the world that bore him. In contradistinction to the fortified walls of Utopia, his trajectory is one of self-destruction. He fights to defend his world, not for the value it merits, but for the absence of ethical valuation it enables. The creed of this anti-hero is 'oblivion now'. As he states plainly in the novel's opening pages: 'I don't believe in the existence of anything at all.'[43]

The ethical flatness of Tawfiq's utopians is mirrored by the author's vision of the dystopic world that surrounds them. This becomes evident midway through *Utopia* when the nameless protagonist and his female side-kick Germinal find themselves trapped in the city of Cairo, also known as the territory of the 'Others'. Their guide, and captor, Gaber, leads them on a tour of the decrepit streets, negotiating traffic and crowds inundated with men 'ready to fight for any excuse'. The hostages, inhabitants of an adjacent society known as 'Utopia' (where dogs are 'groomed for protection, not eating'), move timidly through the wasteland. Sensing fear Gaber remarks: 'it's the morality of crowds. Put six chickens in a coup and see how well they behave. See if one doesn't pluck out the eyes of all the others.'[44] Still, the desperate population multiplies, he explains, despite efforts on the part of the 'rulers' to 'castrate' the men or to poison the food with a 'contraceptive gossypol'.[45] 'Why don't you revolt?' The narrator asks:

> That's batted around now and then. But the twentieth century is over and the revolutions that used to quell the masses are long gone. Those at the top have learned from the mistakes of their predecessors. You'll never see another Iranian Shah circling around in his helicopter looking for a place to land, or the corpse of a Ceausescu, or Mussolini hanging in the public square. The security apparatus today has evolved. There are six security systems monitoring each other all for the single purpose of protecting the rulers. Revolutions today are closer to riots. The choppers appear, fire some shots, and everybody scatters.[46]

The speaker's cynicism towards the act of revolt becomes ironical in the end. His murder and the rape of his sister by the narrator-protagonist create the impetus for the final scene of the novel that finds a mass of humanity from the territory of the 'Others' marching on the gated realm of 'Utopia'. Striking a motif Tawfiq would continue to revisit, the narrator himself casts the final blow, firing a machine gun indiscriminately towards the approaching crowd.

Cinematic in style and tied closely to the global currency of speculative fiction, fantasy and horror, *Utopia* intersected with the kind of critical dystopic reflections on late capitalism that have been endemic to world literature for over a century, from Huxley's *Brave New World* (1931), to Orwell's *1984* (1949), or Gibson's *Neuromancer* (1984). Debauchery begets annihilation in *Utopia* – an idea and an image closely akin to the project of succession surrounding the Mubarak clan,[47] on the one hand and, on the other, a more timeless narrative of Western decadence: 'Utopia' in the year 2023 is guarded by none other than the US Marine Corps.

But the narrative trajectory set in motion by this novel has been evolving in Tawfiq's subsequent works, becoming in essence a veritable reservoir of 25 January angst. Flash forward to the past: Tawfiq's next novel, *al-Sinja* (The knife, 2012), returns to a world similar to that of the Others. Only now the surge of revolution is upon us and the youth of 'Dahdira al-Shanawi', the wretched quarter on the outskirts of the city where the novel is set, move back and forth between Tahrir and the *hara* (the slum), sleeping some nights in the square and intermingling with the 'middle class kids' ('with posters of Che Guevara on their walls and an aptitude for computers') who 'sparked' the uprising.[48]

News of the revolution arrives largely through hearsay. The narrative perspective of the novel remains limited to the confines of 'Dahdira al-Shanawi' (the name itself suggests a hole in the ground) where the narrator-detective is working to piece together clues behind the death of a young girl. But mysteriously – or perhaps conveniently – the two storylines are connected. Returning to his roots as a young adult fiction writer, Tawfiq's *al-Sinja* is a modern day ghost story. Like *Utopia*, however, it also aspires for social allegory. Afaf's corpse, found along the tracks of the 'iron dinosaur . . . no longer capable of frightening the children', is a metonym for the people – *al-sha 'b*.[49] 'Just when the tyrant thinks the body he has pummeled is dead', the narrator thinks, 'that is the moment when it rises again to seek its revenge.'[50] For the narrator of *al-Sinja*, the girl's death is intertwined with the symbols and graffiti he finds peppering the walls of Dahdira. Still her message seems hopelessly ambiguous. The most important clue is a word found sprayed against a wall near the train tracks where her body is found: '*al-sinja*' (the knife). Or was it *al-sibha* (the majesty of God)? Or *al-sija* (Sija)?[51] Or *al-siranja* (the syringe)? Or *al-sirja* (hopscotch)? 'Everything merges', he laments.[52]

Messages from beyond the dead are seldom transparent. But they have tremendous power in convening the living. As the narrator imagines the dead girl – the 'young, fresh face that seemed to have no connection with his own generation' – he sees assembled in Tahrir the youth of the quarter from which she was snatched. The pages populate with these youthful voices and as he witnesses the crowds amass in Tahrir, 'over cellphones', the narrator wonders: 'maybe Nawal is there, maybe Hussein, maybe Jamal al-Faqi, maybe Alaa'.[53] Reminiscent of the real mobilisation of Egyptian youth mythically connected by the post-mortem image of Khaled Said, whose shattered face was shared by tens of thousands on Facebook, the outpouring of rage Tawfiq paints is an expression of life. But this 'picture of a thousand dots blurs to become an awesome giant'.[54] He recalls the meaning of Afaf's final word and thinks of the 'white weapons' (*aslaha bayda*) flooding the city's popular quarters. As much a battle cry as it was a warning: 'the price of democracy is high and the revolution will cost us'. 'Even civil war', he predicts, may fail to settle the score.[55]

The Millennials

Tawfiq is a *sui generis* writer and his path to literary stardom has been an unlikely one. A medical doctor by training, he began writing for the youth fiction market in the late eighties. By the time he turned to novel-writing with *Utopia*, he had published some 500 short works, dime-store novellas

Figure 8.1   White weapons in Cairo, 29 January 2011. © Nathaniel Greenberg

most of which could be characterised by a description he coined: *adab al-ru'b* or 'Horror fiction'.[56] Apart from his translations, which include George Orwell's *Nineteen Eighty-Four* and a 2005 translation of Chuck Palahniuk's *Fight Club* (1996), his most famous work, *Ma wara' al-tabi'a* (The supernatural, 1992–), is a detective series featuring a protagonist, 'Raf'at Isma'il', who travels the globe – mainly Europe – in search of ancient and 'mysterious legends'. Tawfiq likes to claim, for example, that his 'legend of the vampire' (*Massas al-dima'*) from *Ma wara' al-tabi'a* was the first to introduce Egyptian readers to the character of Dracula.[57]

The publishing industry can hardly keep up with Tawfiq and certainly translations (apart from Chip Rossetti's *Utopia*) have lagged behind. All four of his novels to date have topped the bestseller list of Egypt's largest publisher (and the parent company of the eponymous newspaper) Dar al-Shorouk. (*Utopia* has been in the top twenty for over half a decade.)[58] Like his predecessor and early mentor Nabil Farouk, Tawfiq produces fiction at a breakneck pace. Along with *Ma wara' al-tabi'a*, the series *Safari* (1996–) about a medical explorer in Sub-Saharan Africa, and *Fantasiya* (1995–), based on the American television show *Quantum Leap*, each have over fifty instalments to date. But, to be certain, the popularity of Tawfiq's literature is grounded in the demographics of his readership. 'Frightening tales never fail to attract children', he said in an interview for the campus press at Ain Shams University in the Abbassiya neighbourhood of Cairo ('Raf'at Isma'il is a role model for so many young people!' exclaimed his interviewer).[59] This is significant as a larger percentage of the youth demographic in Egypt is literate compared to their older counterparts – 92 per cent and 84 per cent for boys and girls aged 15–24,[60] compared to 82 per cent and 65 per cent of the adult population.[61] As pointed to in a recent study by the publishing house Dar al-Ruwaq, readers under the age of twenty-three constitute 80 per cent of the young fiction market that Tawfiq's literature serves.[62] Moreover, the youth population is not only the most literate segment of the population in Egypt but it is also the largest. At the time of the 2011 uprising, more than half the population was under the age of twenty-five.[63]

In addition to demographics, Tawfiq's work came to occupy a certain vanguard position in a now booming genre in speculative and science fiction (*al-khayal al-'ilmi*). From digital comic books, to movies, music videos, and video-gaming, the Arab world has witnessed a surge in this genre over the past decade. Literature has been at the forefront of the wave with writers like Hassan Blasim (Iraq), Ahmed Sadawi (Iraq), Nail al-Tukhi (Kuwait), Boualem Sansal (Algeria), Fadi Zaghmut (Jordan), Bassima Abd al-Aziz (Egypt) and Mohammad Rabie all receiving major international acclaim for their novels in recent years. In addition to having written the largest grossing such novel with *Utopia*, Tawfiq, along with Sadawi who won the Arab Booker for his novel in 2014, also garnered one of the industry's most coveted awards winning the Sharjah International Book Award for *Mithl Ikarus* (Like Icarus) in 2016.

Black Box

Most notable about this development is the basic observation, as Tawfiq wrote in a 2010 article, that the 'Arab literary movement had never taken science fiction seriously nor had the latter ever assumed the form of a *tayyar* [or trend]'. That is, independent of a few isolated individuals, or works, there existed no genre, *per se*. Tawfiq posited two theories for understanding this void: (1) the Arabic literary milieu was spacious such that writers felt little need to pursue more specialised fields, e.g. science fiction, police drama, spy fiction (to each of which there are exceptions), and (2) (less optimistically) science fiction was 'born of a science consuming culture' and so, *res ipsa loquituir*, the Arab experiment remained stillborn.[64]

In recent years Western scholars have sought to reconstitute the literary history of Arabic science fiction by digging up relics and tying together various disparate strands of Arab poetics, from legends, to fables, even *1001 Nights*. Ada Barbara's recent book, *La fantascienza nella letteratura araba* is the most illustrative such effort. Among other things, Barbara pointed to the millennial-old tradition of *'aja'ib* or 'marvels' as it manifest in post-war Arabic sci-fi. 'Proto-fantascienza' examples like Ibn Tufayl's epistle of mystical isolationism *Hayy Ibn Yaqzan*, or folkloric elements of the medieval *siyar* appeared equally suggestive when read through a historicist lens. But it was also easy to overstate the legacy of this otherwise nascent genre. Literary scholars indebted to the schooling of Derridean poststructuralism tend to be loath to admit contextual reckonings, particularly when the linkage appears obvious. There was clear coincidence between the current surge in Arab sci-fi writing and the utopic imaginings of the past half-decade, from Tahrir to Taizz, Raqqa and Mosul. The dynamic was arbitrary of course and certainly there were other socio-historic fusion points that made sense – the neoliberal market reforms of the 1990s, or the penetration of ICT in the early twenty-first century. Nonetheless, as most lay scholarship on Arabic sci-fi has shown, the subordinating of contemporary cultural phenomena to the exotica of an imagined past shows little sign of abating. Described by scholars like Salama Musa and Abbas Mahmoud al-Aqqad as early as the 1930s as a way of deflecting through the illusion of a continuum the actual fragmentation of culture during the decades of decolonisation,[65] the historicisation of Arabic sci-fi has tended to surface mostly among lay critics in the West.[66] While commenters writing in Arabic were not immune to the impulse, the prevailing theory on the rise of Arabic sci-fi in Arab media reflected closely the theoretical binary expressed by Tawfiq.[67]

In the quest for origins, as Tawfiq and others like Yusuf Sharuni have shown, a more material point of departure existed in the pioneering experimentalism of the early twentieth century. Salama Musa's collection of short-stories, *Ahlam al-filasafa* (The philosopher's dream, 1926) addressed, among other things, the evolution of utopic thought from the time of Plato through the industrial revolution and onward to Egypt in the year 3015. Tawfiq al-Hakim's short-story, *Fi sanat milyun* (In the year one million, 1950) and play, *Rihla ila al-ghad* (Journey to the future, 1958) – though veering closer to the kind of neo-psychologism of Abbas Mahmoud al-Aqqad, or perhaps, as Barbaro suggested, George Orwell, than a Phillip K. Dick or even Jules Verne – is credited often with sparking popular interest in science fiction. And Naguib Mahfouz, whose neo-*maqama Malhamat al-harafish* (The harafish, 1977) was steeped in the tradition of '*aja'ib*, created, in 1959, something of a prototypical sci-fi hero with the fifth and final protagonist of his post-revolutionary masterpiece *Awlad haratina* (Children of the alley, 1959): ''Arafa', or the 'man of science'. In many ways, pillars of modern Arabic literature writ large, these writers fashioned into the core of their aesthetics a critical positionality on science that – extending from the experience of the *Nahdha*, or 'renaissance' of the late nineteenth century – was at once divorced from the cultural logic of the West but sharply critical still of religious positions that refuted as a point of social contention the ethical validity of scientific inquiry. Mahfouz's Arafa, for example, in his attempt 'to know' 'Gabalawi' ultimately murders the 'unseen' overlord of the alley.[68] For his crime he is captured by the guardians of the 'big house', the *futuwwat*, who bury him alive. Arafa's 'science', a book of 'magic', is smuggled away by a compatriot, salvaged we imagine, for future generations.[69]

The first explicit iteration of science fiction in Nihad Sharif's *Qahir al-zaman* (Time conqueror, 1972) begins where Arafa left off. 'Was he heading to the mountain or to the villa to discover its secrets?' wonders the narrator about the mysterious mansion in the Muqattam Hills where five years prior an employee was found murdered.[70] Much like Mahfouz's *Children of the Alley*, 'Time conqueror' centres on the intrigue of a walled enclave in Old Cairo. Here the residency is occupied not by an ageless deity figure but a scientist – Doctor Halim Sabru – who has created a 'time travel' machine. Written as a modern adaptation of H. G. Wells's *The Time Machine* (1895), *Qahir al-zaman* inverts the science at the centre of the English classic, interconnecting through the character of Dr Sabru an element of timelessness – or uchronia in sci-fi lingo – that resonates more with Mahfouz's ageless Gabalawi than Well's time traveller. Rather than travel through time and back, Dr Sabru's machine allows him to live indefinitely, frozen cryogenically, while history, as it were, moves on. (Tawfiq takes up this mantle directly. The only English to grace the pages of *al-Sinja* is the phrase 'time sharing', an ambiguous allusion to the hyper-communicative lifestyle of millennials. And *Mithl Ikarus*, echoing his series *Fantasiya*, is something of a Wellsian swan song for the Egyptian mind.)[71]

In a broader sense, this vision of immortality, 'a characteristic exclusive to God', as Barbaro notes, would become engrained in the genre, born of hibernation, elixirs, reincarnation, molecular cell operations.[72] In some respects it was a motif central to the very impulse of science fiction. As Fredric Jameson observed: 'in SF . . . religion is the black box in which infrastructure and superstructure mysteriously intermingle'.[73] But in the post-revolutionary context of Mahfouz's *Children of the Alley*, Sharif's 'Time conqueror' or Gamal al-Ghitani's *Waqa'i' harat al-Za'farani* (Incidents in Zafran Alley, 1976), where a travelling salesman peddles a proto-Viagra potion to the residents of old Cairo, literature posited on the idea of immortality becomes as much satirical as it is speculative. From its origin, Nasserite socialism functioned as a sort of 'black box' ideology intermingling and celebrating an 'enigmatic identity' that was equally founded on the material restructuring of modes of production and cultural identity.[74] Mahfouz was one of the first to ironise the identity-based politics of the ideology but his critique was far more subdued than other polemical writers of the time. As the Marxist critic Ghali Shukri wrote, Sadat's invocation of Nasserite socialism was equally disingenuous as his 'vulgar attempt to win the people's confidence . . . by means of religion'.

> This religious-based socialism consists of the verses from the Koran which the authors of the guide use to support their doctrine, and which we often saw pinned to the wall behind the desks of Egyptian capitalists before the revolution.[75]

Islamists as well latched on to the hypocrisy of religious-based socialism. The literary critic and proto-Islamist ideologue Sayyid Qutb famously denounced Nasserite socialism in his 1964 book *Ma'alim fi al-tariq* (Milestones) as a reincarnation of the *jahiliyya*, an age of ignorance experienced by Arabs before the appearance of Islam. Even attempts to reverse the secular leanings of the revolutionary regimes – such as that of the Al-Azhar-trained Haouri Boumediene who inserted himself as President of Algeria after a coup in 1965 – failed to assuage the scorn of detractors. In response to Boumediene's policy of 'Islamic socialism', the prominent Algerian shaykh Abd al-Latif al-Soltani published in 1974 *al-Mazdakiyya hiya 'asal al-ishtirakiyya* (Mazdakism is the source of socialism), a scathing critique that compared the religious-based socialism of Sadat and Boumediene to a mythical tyrant ('Mazdak') from fifth-century Persia – 'land of the barbarians'. Al-Mazdak, Soltani wrote, promised security and reform by 'obligating all of his followers to abide by his principles . . . Such that there existed no difference between them, no unique devotion between brothers, no distinction between the sacred and the profane'.[76]

No less dystopian, Sharif's 'Time conqueror' from 1972 is ultimately a critique of the 'black box' ideologies emanating out of the secular Arab capitals of the post-independence years. Dr Sabru's attempt to master the enigma of mortality ends ultimately in his being crushed to death after his assistant destroys his laboratory in a final act of moral clarity. Like Mahfouz's

Children of the Alley, however, the doctor's notebook titled 'The Living Cell' is smuggled to safety by the dead doctor's butler. But 'Time conqueror' – nor any work of Arab sci-fi for that matter – never faced the kind of censorship enjoyed by the realists, like Mahfouz, or polemicists, like Shukri, Qutb or Soltani; perhaps because Dr Sabru's flirtation with immortality implies not an ideological or political critique but a civilisational one. It is not the 'superstructure' that betrays him, that is, the romance of an afterlife, but the 'infrastructure' of a cryogenically preserved life. It is the science – and by extension the aspirations of Western rationalism – that become the seed of chaos in Sharif's novel. This critique has persisted in Arab sci-fi and is particularly evident in the most celebrated sci-fi work in recent years: Ahmed Khaled Tawfiq's *Mithl Ikarus*.

Pandora's Box

Set in 2020 just five years after its date of publication, and three years before *Utopia*, *Mithl Ikarus* tests the viable limits of social allegory. The reader is torn between a 'willful suspension of disbelief' inherent to the novel's fantastical premise, on one hand, and the inevitable historical innuendo that drifts through its pages on the other. This latter dimension is compounded by the 'science' of the novel: a mysterious archive and hapless soothsayer, 'Mahmoud al-Simnudi', who gains access to an instantaneous and universal knowledge of the world. With the historical events of 2011 in rearview, the 'elixir' of the protagonist's condition serves to illuminate the obvious, but materially elusive notion that the Arab uprisings were preceded by and inherent to the technological revolution of another more ubiquitous archive of instantaneous and universal knowledge, namely, the Internet. The fate of the protagonist and the ultimate triumph of the narrator-antagonist who silences him serve to articulate a critical counternarrative to that of the 'wired revolutionaries': that the science of ICT had reached too far, that its avatars had flown too close to the sun and so were bound to fall.

While the fictional time of the novel – moving from 2020, to the past, to the future and back again to the present – follows in the tradition of Wells, the temporal frame of reference for *Mithl Ikarus* intersects with a narrative of modernity and modernisation that extends beyond the 2011 uprising. At the same time, however, the novel provides a powerful vantage point onto the early weeks of 2011 in part because it helps to explain some of the preconceptions guiding the demonisation of technology following the WikiLeaks disclosures on 27 January that elements of the April 6 group had received funding and training in Washington. As discussed in Chapters 1 and 2 derision towards the so-called youth activists became a common trope in public discourse, not simply because of their association with Western interests – though this was significant – but also, as Tawfiq's novel explores, because communication technology itself by 2011 was beginning to represent a kind of 'Pandora's box', material evidence of Abd al-Fattah al-Sisi's 'Fourth Generation Warfare', the mysterious media campaign to undermine the security of the state.[77]

Still, the novel is less raw than *al-Sinja* which, in an explicit way, diverts sharply to integrate the events of 25 January. In *Mithl Ikarus* there is no singular device, metonymic or otherwise, that serves to invoke the novel's embedded critique. Rather, Tawfiq elevates the plight of his protagonist, a most unlikely candidate for becoming an ''arif', a 'seer'.[78] A rural lawyer from a 'middle class' background with a 'middle class wife' and a 'medium sized house' with 'medium sized furniture', Mahmoud – 'the strange object' – should have been destined 'to die alone like a dog' remarks the narrator.[79] He has no use for infinite knowledge; could not care less about the mysterious 'Askashic records', or those before him who used their power to 'decapitate the consciousness of humanity'.[80] In this way, Mahmoud al-Simnudi is an inversion of the Mahfouzian 'man of science'. His magic is a burden, not a gift. This ironical turn adds welcome comedy to the story. 'This man knows everything' says an American 'General' seated with a panel of experts assembled to ply Mahmoud for information. 'He knows our secret sources. He predicted our attempt to assassinate the dictator . . . Our man . . . The one we betrayed . . . This man, Mahmoud, he is the real thing and we have to know what he knows.' One of the Americans on the panel then turns and asks: 'does he speak English?' To which the General replies: 'He uses it brilliantly but its gobbledygook. You won't understand him.'[81] The early invocation of Kubrick's *Dr. Strangelove* forecasts the tonality Tawfiq is aiming for with the cartoonish Americans that appear intermittently throughout the novel. But it is the work's wry if feverish effort to embellish the events of 25 January as part of an American conspiracy designed, as Mubarak famously proclaimed in his midnight speech on 28 January, 'to destabilize the establishment and undermine the law',[82] that most clearly drives the novel's prevailing cynicism towards the value of mass information and the intention of those who would wield it.

The dramatic final monologue by Mahmud al-Simnudi's interrogator captures the novel's counter-revolutionary spirit. 'I am the hero who conceals once more the truth so that no one will witness its nakedness', he says after torturing and ultimately killing al-Simnudi. 'I am the hero that restores for the people their will to live another day.'[83] It is an inevitable and strangely welcome ending for Mahmud. Over the course of the novel the character endures an unending series of interrogations in Egypt as well as the US where an anonymous group of scientists and security experts ply his mind for knowledge of an impending disaster. But like Jorge Luis Borges's famous story *Funes el memorioso* (1942) about a man who, after being kicked in the head by a horse, suffers from infinite memory and is forced to live in silence and darkness as a result, Mahmoud masters the art of 'complete sensory deprivation' (p. 97). He does this to execute his mysterious method of mental time travel, but also to silence himself to the present. The practice renders him something of a 'sleeping prophet',[84] the narrator explains, a description that, like many in the novel, implies martyrdom is not far behind.

As the events of 25 January unfolded, Tawfiq kept a close eye on the ostensibly destabilising dimension of technology in the struggle for Tahrir. 'How strange', he wrote in a blog post on 3 February 2011:

> the idea of our thugs hurling torrents of Molotov cocktails on our demonstrators while viewers around the world sit by and watch, eating popcorn and drinking cola. The corpses of horses and humans mount atop one another. Mosques convert to field hospitals. Somehow the museum still stands, narrowly averting cosmic catastrophe. Is this really Egypt? Who did this? Who transformed this beautiful symphony of youth into a scene of such carnage? Certainly not the youth, as our media would have it. It is not Adil the gamer, or Maha the blogger. One must be insane to believe that it was Mohsen the telecommunications engineer wielding bottles of flaming water. Let us remember that Maha is bleeding, that Adil has been martyred, and that Mohsen has been disappeared.[85]

The illusory effect of vast online networks, present in mind but absent in reality, soon manifest in *al-Sinja* where the revolution (like the dead girl's dispatches) 'resembles a game Westerners call "Chinese telephone"'.[86] The more the message multiplies the less coherent it becomes. In a less explicit way Tawfiq's textured, if graphic prose also serves to divert the reader's gaze to the opposite extreme of the sensorial spectrum. 'The smell of drawn blood reminds you of something' *Muhl Ikaraus* begins. 'Something I knew in the past perhaps; some prehistoric memory of the cave; some past life as a killer, thirsting for blood.'[87] As with the wild-eyed spectacle of Tawfiq's narrator-protagonists 'killing to have hunted',[88] the novels' playful interchangeability of narrative perspective (shifting from first, to second, to third person omniscient) complements the author's relentless focus on the sheer carnality of social existence. More visceral still (and the effect most readily imitated by younger admirers like Mohammad Rabie) are the visual iterations of this illocution: the affectless severing of body parts (*Utopia*), the flesh-eroding slums (*Utopia*, *al-Sinja*), 'the body in the center of a field soaked in blood' and the killer who streaks out of view.[89]

Flattening, expanding and yet, at once, dramatically shrinking the social universe, digital communication technology in general infused the greater body of contemporary Arab sci-fi with an ever-more 'disembodied poetics'. Perhaps closer to a nominal sense of the term '*khayal*', which refers literally to ghosts, one finds in Tawfiq's brutal narrators a prevailing sense of ambiguity towards the corporeality of human existence, a vision of human life detached from and yet haunted by the very spirit the body no longer recognises. This literary numbness towards life preceded the great Arab uprisings, which seemed to synchronise through experience – on streets and sidewalks – that which had been severed by the illusion of connectivity. Critically, however, these speculative fictions also served to frame events as they unfolded. In a March 2011 interview with the magazine *Kelmetna*, Mahmoud Uthman, whose 2007 cult classic *Thawrat 2053* has been described as 'prophetic' of the 2011 uprising, said of the mysterious character at the centre of his novel who projects the future through photographic images but refuses to carry a cellphone, that he was not 'hostile to technology' *per se*, but hostile to technologies of 'human communication'. He adds: 'the cellphone in my opinion has destroyed much of what is precious in society, so that's how I determined his [the Stranger's] position on technology.'[90]

Conclusion

Writers like Tawfiq or Uthman had been tapped into popular anxieties surrounding ICT well before the outburst of the Arab Spring. Many of the speculative works of fiction now peppering the Arab cultural landscape underscore an impulse to rewrite history. Others, like Magdy al-Shafee's *Metro* (2008), celebrated it as a vehicle of empowerment. The graphic novel begins with its hero, a software designer, mass dialing passengers waiting for a train at an underground station – 'Mohammed Naguib' – named for the country's first president following independence in 1952.

Invariably dystopic, the genre of Arabic sci-fi emerged as a reflection of the revolutionary trajectory once taken to its logical extreme. From the limits of social cohesion to the cohesion of self, such fiction paints the parameters of a rapidly changing world.

Tawfiq's 2014 graphic novel *Ta'thir al-jarada* (drawings by Ahmed Atif Mujahid) satirised that experience. Drawing on the tradition of Sharif's *Qahir al-zaman* and forecasting what he would do with *Mithl Ikarus*, *Ta'thir al-jarada* featured a time-travelling scientist who is summoned to the bunkered residence of a recently ousted President Hosni Mubarak and his sons. For a price, the time traveller, an unassuming middle-aged man who wears glasses and is balding, agrees to travel back in time to reverse the course of history. His stops include the Battle of the Camels, the morning of 25 January, the parliamentary elections of 2010, Vienna and the office of Mohamed ElBaradei who called for a boycott of the elections that September, and finally the Internet café in Alexandria on the night Khaled Said was killed, 6 June 2010. This final stop, however, is his last. Flash forward to 2011, Mohamed Bouazizi has died, the Tunisians have revolted and the tyrant has fled. Protests are raging in the streets of Cairo and history is on track but for one small difference. The 'We are all Khaled Said' Facebook page has been edited. In attempting to intervene in the murder of the young blogger the time traveller was also killed. And the page now reads: 'We are all Khaled Said and that other guy we don't know'. History has been altered but the result is the same. The revolution rages regardless.

Notes

Introduction

1. Chapter 5 unpacks the genealogy of this description, including the erroneous report that Bouazizi was a 'college graduate'.
2. Adunis, 2011, 'Ramad al-Buʿazizi', *Al-Arabiya*, 28 April, https://www.alara biya.net/views/2011/04/28/147037.html (accessed 24 February 2018).
3. Ibid.
4. Fredric Jameson, 1971, *Marxism and Form*, Princeton: Princeton University Press, p. 259.
5. Paul Ricoeur, 2005, *The Course of Recognition*, Cambridge, MA: Harvard University Press, p. 100.
6. Ibid. p. 100.
7. Bruce Lincoln, 2012, *Gods and Demons, Priests and Scholars: Critical Explorations in the History of Religions*, Chicago: University of Chicago Press, p. 55.
8. The White House, 2011, 'Remarks by the President on the Middle East and North Africa', The White House Archives, 19 May, https://obamawhitehouse. archives.gov/the-press-office/2011/05/19/remarks-president-middle-east-and-north-africa (accessed 28 February 2018).
9. Al-Arabiya Net, 2013, 'Al-Ghannouchi: Belʿid laysa al-Buʿazizi wa ana laysa Ben ʿAli', *Al-Arabiya*, 10 February, http://www.alarabiya.net/artic les/2013/02/10/265374.html (accessed 3 October 2017).
10. Sylvain Attal, 2011, 'Tahar Ben Jelloun: Écrivain', *France 24*, 14 July, http:// www.france24.com/fr/20110713-tahar-ben-jelloun-etincelle-par-le-feu-revolution-arabe-revolte-vengeur-tunisie-tunis-bouazizi (accessed 24 February 2018).
11. Marwan M. Kraidy, 2016, *The Naked Blogger of Cairo: Creative Insurgency in the Arab World*, Cambridge, MA: Harvard University Press, p. 30.

12. Walter Benjamin, 1955, 'The Storyteller', in W. Benjamin, *Illuminations*, New York: Schocken Books, p. 89.

13. See Hans Robert Jauss, 1982, *Toward an Aesthetic of Reception*, trans. Timothy Bahti, intro. Paul De Man, Minneapolis: University of Minnesota Press.

14. Wael Ghonim, 2013, *Revolution 2.0: the Power of the People is Greater Than the People in Power: a Memoir*, Boston: Mariner Books/Houghton Mifflin Harcourt, p. 59.

15. Quoted in Linda Herrera, 2014, *Revolution in the Age of Social Media*, London: Verso, p. 55.

16. Ibid. p. 64.

17. Ibrahim Abduh, 1964, *Jaridat Al-Ahram: tarikh wa al-fann 1875–1964*, al-Qahira: Mu'assasat Sijill al-'Arab, p. 23.

18. B. Turck, 1972, 'The authoritative Al-Ahram', *Aramco World* 23.5, http://archive.aramcoworld.com/issue/197205/the.authoritative.al-ahram.htm.

19. A video of the performance is available at: http://www.anazahra.com/entertainment/photo-126976/%D8%AA%D9%81%D8%AC%D9%8A%D8%B1-%D8%A7%D9%84%D8%A7%D8%B3%D9%83%D9%86%D8%AF%D8%B1%D9%8A%D8%A9-%D9%8A%D8%B4%D8%B9%D9%84-%D8%AD%D9%85%D9%89-%D8%A7%D9%84%D8%A3%D8%BA%D8%A7%D9%86%D9%8A-%D8%A7%D9%84%D9%88/.

20. Tarek Osman, 2011, *Egypt on the Brink*, New Haven: Yale University Press, p. 210.

21. Said, Mekkawi, and Amr Kafrawi, 2014, *Kurrasat al-Tahrir: hikayat wa-amkina*, Cairo: Dar al-Misri al-Libnaniya, p. 134. On Musa's historic speech, see Al-Arabiya, 2011, ''Amr Musa . . .', *Al-Arabiya*, 11 February, http://www.alarabiya.net/articles/2011/02/11/137244.html (accessed 1 February 2017).

22. See Emarat al-youm, 2011, ''Omar Suleiman Na'iban l-il Mubarak . . . wa Shawfiq ra'isan', *Emarat al-youm*, 30 January http://www.emaratalyoum.com/politics/news/2011-01-30-1.348735 (accessed 24 February 2018).

23. Said, 2014, p. 53.

24. David Osnos *et al.*, 2017, 'The new Cold War', *The New Yorker*, 6 March, https://www.newyorker.com/magazine/2017/03/06/trump-putin-and-the-new-cold-war (accessed 24 February 2018).

25. Plato, 1992, *The Republic*, trans. G. M. A. Grube and C. D. C. Reeve, Indianapolis: Hackett Publishing Co., p. 232.

26. Ricoeur, 2005, pp. 101–2.

27. Youssef Seddik, 2012, 'La chronique de Youssef Seddik: Encore une lettre à un(e) inconnu(e) intelligent(e) parmi nos gouvernants provisoires', *Le Temps*, 25 August, http://www.letemps.com.tn/article-69172.html (accessed 1 July 2013).

28. Jürgen Habermas, 1990, *The Philosophical Discourse of Modernity: Twelve Lectures*, Cambridge, MA: MIT Press, p. 286.

Chapter 1

1. Solomon E. Asch, 1952, *Social Psychology*, Englewood Cliffs: Prentice Hall Inc., p. 454.

2. Ibid. p. 457.

3. Researchers Sander van der Linden, Anthony Leiserowitz, Seth Rosenthal and Edward Maibach draw the phrase 'merchants of doubt' from a 2011 book, *Merchants of Doubt*, by N. Oreskes and E. M. Conway. See van der Linden *et al.*, 2017, 'Inoculating the public against misinformation about climate change', *Global Challenges* 1.2.

4. Ibid.

5. Ibid. See also BBC, 2017, 'Cambridge scientists consider fake news "vaccine"', BBC, 23 January, http://www.bbc.com/news/uk-38714404 (accessed 30 June 2017).

6. The phrase 'the Asch situation' derives from the name of a study László cites. L. Ross, G. Bierbauer and S. Hoffman, 1976, 'The role of attribution processes in conformity and dissent: revisiting the Asch situation', *American Psychologist* 31, 148–57.

7. János László, 2008, *The Science of Stories: An Introduction to Narrative Psychology*, New York: Routledge, p. 59.

8. The World Bank, 2017b, 'Mobile cellular subscriptions (per 100 people)', http://data.worldbank.org/indicator/IT.CEL.SETS.P2?end=2015&locations=EG&start=1999 (accessed 3 June 2017).

9. The World Bank, 2017a, 'Individuals using the Internet (percent of population)', https://data.worldbank.org/indicator/IT.NET.USER.ZS?end=2015&locations=EG&start=1999 (accessed 3 June 2017).

10. Paul Ricoeur, 2005, *The Course of Recognition*, Cambridge, MA: Harvard University Press, p. 98.

11. Ibid. p. 98.

12. Jürgen Habermas, 1990, *The Philosophical Discourse of Modernity: Twelve Lectures*, Cambridge, MA: MIT Press, p. 286.

13. The phrase 'art as experience' was coined by John Dewey with his book of the same title in 1934.

14. See, for example, N. Greenberg, 2013, 'Emergent public discourse and the constitutional debate in Tunisia: a critical narrative analysis', *TelosScope*, 30 December, http://www.telospress.com/emergent-public-discourse-and-the-constitutional-debate-in-tunisia-a-critical-narrative-analysis/ (accessed 27 September 2017).

15. László, 2008, p. 67.

16. Ibid. p. 67.

17. See Nathaniel Greenberg, 2011, 'A people's protest?', *The Seattle Times*, 28 January, https://www.seattletimes.com/nation-world/a-peoples-protest-the-view-from-a-cairo-coffeehouse/ (accessed 24 February 2018).

18. Hans-Georg Gadamer, 1975, *Truth and Method*, New York: Continuum, p. 34.

19. Tim Lister, 2011, 'U.S. cables: Mubarak still a vital ally', *CNN*, 28 January, http://www.cnn.com/2011/WORLD/africa/01/28/egypt.wikileaks.cables/ (accessed 11 October 2017).

20. Mark Landler and Andrew W. Lehren, 2011, 'Cables show delicate U.S. dealings with Egypt's leaders', *The New York Times*, 27 January, http://www.nytimes.com/2011/01/28/world/middleeast/28diplo.html?_r=0 (accessed 11 October 2017).

21. Margaret Scobey, personal interview with the author, 1 November 2016.

22. Tim Ross, Mathew Moore and Steven Swinford, 2011, 'Egypt protests: America's secret backing for rebel leaders behind uprising', *The Daily Telegraph*, 28 January, http://www.telegraph.co.uk/news/worldnews/africaandindian ocean/egypt/8289686/Egypt-protests-Americas-secret-backing-for-rebel-leaders-behind-uprising.html (accessed 11 October 2017).

23. See, for example, David Filipov, 2017, 'The notorious Kremlin-linked troll farm and the Russians trying to take it down', 8 October, https://www.washingtonpost.com/world/asia_pacific/the-notorious-kremlin-linked-troll-farm-and-the-russians-trying-to-take-it-down/2017/10/06/c8c4b160-a919-11e7-9a98-07140d2eed02_story.html?utm_term=.0989b15174e0 (accessed 11 October 2017).

24. Ross *et al.*, 2011. In 2017, I contacted Tim Ross to inquire about the backstory to his article. Three months later (for reasons perhaps unrelated) the comments section from the article had been removed. I saved much of it however using screenshots.

25. 'Tropicgirl' was not the only user from the 27 January *Telegraph* article to reappear in support of Trump and far-right narratives. The author has retained the names of at least half-a-dozen.

26. See, for example, Nicholas Jackson, 2011, 'WikiLeaks' Assange: Facebook is an appalling spy machine', *The Atlantic*, 2 May, https://www.theatlantic.com/technology/archive/2011/05/wikileaks-assange-facebook-is-appalling-spying-machine/238225/ (accessed 24 February 2018).

27. See Emma L. Briant, 2015, *Propaganda and Counter-Terrorism: Strategies for Global Change*, Manchester: Manchester University Press, pp. 90–1.

28. See Policy Coordinating Committee on Public Diplomacy and Strategic Communication (US), 2007, *U.S. National Strategy for Public Diplomacy and Strategic Communication*, pp. 4–5, http://www.au.af.mil/au/awc/awcgate/state/natstrat_strat_comm.pdf.

29. Ibid. pp. 4–5.

30. Linda Herrera, 2014, *Revolution in the Age of Social Media*, London: Verso, p. 38.

31. Howcast Media, 2009, 'How to Smart Mob', YouTube, https://www.youtube.com/watch?v=2prHm1BcU1x (accessed 6 July 2018).

32. Herrera, 2014, p. 38.

33. Julian Assange, 2016, *When Google Met WikiLeaks*, London: OR Books, p. 149.

34. Salamander Davoudi and Ben Fenton, 2011, 'Assange signs deal with U.K. Telegraph', *The Financial Times*, 31 January, https://www.ft.com/content/0ff84f8c-2d53-11e0-9b0f-00144feab49a?mhq5j=e6 (accessed 26 October 2017).

35. WikiLeaks it should be said has proven adept at distancing itself, or conversely, claiming credit for its information warfare when the results are seen as politically expedient. For example, Assange has claimed that the Arab Spring was one of his organisations greatest 'successes' (Assange, 2016, p. 149). In the case of Donald J. Trump, however, Assange has been reluctant to claim any credit.

36. Habermas, 1990, p. 178.

37. R. Schmidle *et al.*, 2015, *White Paper on Social and Cognitive Neuroscience Underpinnings of ISIL Behavior and Implications for Strategic Communication, Messaging, and Influence*, May, https://info.publicintelligence.net/SMA-ISIL-MessagingInfluence.pdf (accessed 15 October 2017).

38. Roland Barthes's essay 'La mort de l'auteur' ('The Death of the Author') was published in 1967.

39. Schmidle *et al.*, 2015.

40. H. L. Goodall identifies three major strands of Department of Defense funding for outside research on matters relating to cultural communications and counter-terrorism. Beginning in 2009, these included the controversial Human Terrain System which embedded academics – anthropologists and social scientists – within military units in Iraq and Afghanistan; a fifty-million dollar Minerva Research Initiative; and an aegis of Broad Area Announcement grants from Office of Naval Research titled 'Understanding Human, Social, Cultural, and Behavioral Influences' (H. L. Goodall, 2010, *Counter-narrative: How Progressive Academics can Challenge Extremists and Promote Social Justice*, Walnut Creek: Left Coast Press, p. 99).

41. Habermas, 1990, p. 178.

42. Schmidle *et al.*, 2015, pp. 126–34.

43. Ibid. pp. 126–34.

44. National Commission on Terrorist Attacks upon the United States, 2004, *The 9/11 Commission Report: Final Report of the National Commission on Terrorist Attacks Upon the United State*, New York: Norton, p. 91.

45. Ibid. p. 91.

46. For a good discussion of the (mis)appropriation of narratology by defence-intelligence analysts and think tanks, see Andrew Glazzard, 2017, *Losing the Plot: Narrative, Counter-Narrative, and Violent Extremism*, The Hague: International Centre for Counter-Terrorism, https://icct.nl/wp-content/uploads/2017/05/ICCT-Glazzard-Losing-the-Plot-May-2017.pdf.

47. Habermas, 1990, p. 181.

48. Ibid. p. 189.

49. Ibid. p. 183.

50. James K. Glassman, 2008, 'Public Diplomacy 2.0: a new approach to global engagement', speech delivered at the New America Foundation, 1 December, https://2001-2009.state.gov/r/us/2008/112605.htm (accessed 19 June 2018).

51. Ibid.

52. Lina Khatib, W. Dutton and M. Thelwall, 2012, 'Public Diplomacy 2.0: a case study of the US Digital Outreach Team', *Middle East Journal* 66.3, 453–72.

53. See Arturo Munoz, 2012, *U.S. Military Information Operations in Afghanistan: Effectiveness of Psychological Operations 2001–2010*, Santa Monica: RAND.

54. In her 2015 book *Propaganda and Counter-Terrorism*, Emma Briant details how the outsourcing of propaganda campaigns by the CIA and other government agencies, beginning in the 1980s and continuing through to the

present day, served to multiply the number of voices engaged in advancing US interests while also normalising 'propaganda processes', in effect creating a perception of independence between the vehicles of propaganda (in the private sector as well as academia) and the respective government agencies whose allocations enable such work (Briant, 2015, pp. 90–1). Several such deals made news headlines in the early days of the wars in Iraq and Afghanistan as multi-million dollar contracts were awarded to private public relations and advertising firms for the purpose of influencing public opinion in the warzone countries (Nathaniel Greenberg, 2017, 'Mythical state: aesthetics and counter-aesthetics of the Islamic State in Iraq and Syria', *The Middle East Journal of Culture and Communication* 10, 255–71, http://booksandjournals.bril lonline.com/docserver/journals/18739865/10/2-3/18739865_010_02-03_ s009_text.pdf?expires=1508169166&id=id&accname=id23163&checksum= 2F6435FDC834653FA49FED31FC97D47). This included the wholesale establishment of locally-based media operations such as the Iraqi Media Network (later *Al-Iraqiya*), as well as 'covert "perception management" campaigns' directed towards international media and online networks, part of the 'full-spectrum' approach envisioned by President Obama with the establishment of US Cyber Command in 2009 (Briant, 2015, p. 88).

55. Khatib *et al.*, 2012.
56. See Dennis M. Murphy and James F. White, 2007, *Propaganda: Can a Word Decide a War?* The US Army War College, http://www.dtice.mil/docs/cita tions/ADA486008 (accessed 19 June 2018); also Philip M. Taylor, 2007, *The Projection of Britain: British Overseas Publicity and Propaganda, 1919–1939*, Cambridge: Cambridge University Press, p. 65.
57. Murphy and White, 2007.
58. Edward Said, 2013, 'Introduction' to the Fiftieth Anniversary Edition of Erich Auerbach's *Mimesis: The Representation of Reality in Western Literature*, Princeton: Princeton University Press, p. 12.
59. Ernesto Laclau, 2005, *On Populist Reason*, New York: Verso, p. 83.
60. Ahmed al-Khatib, 2014, 'Kol ma yajib taʿrifuhu ʿan hurub al-jil al-rabiʿ', *Sasapost*, 26 August, http://www.sasapost.com/4th-generation-warfare/ (accessed 21 July 2017).
61. Glassman, 2008.
62. While beyond the scope of this particular book it is worth noting that significant research has been invested in developing precisely this strategy. Of paramount interest is the work of researchers associated with the French

Research Center for Intelligence. Both Yanick Bressan and Philippe-Joseph Salazar have published major works in recent years addressing the neurological dimensions of aesthetic reception, including in regards to ISIS recruitment strategies (Bressan, 2018, *Radicalisation, renseignement et individus toxiques: mieux comprendre les processus de manipulation mentale*, Paris: VA éditions) and the rhetorical science underpinning ISIS communications and counter-communications strategy (Salazar, 2017, *Words are Weapons: Inside ISIS's Rhetoric of Terror*, New Haven: Yale University Press).

63. Raymond Williams, 1986, 'The uses of cultural theory', *The New Left Review* I.158, https://newleftreview.org/I/158/raymond-williams-the-uses-of-cultural-theory (accessed 28 February 2018).

64. See the US Department of Justice, 2018, 'Internet Research Agency indictment', https://www.justice.gov/file/1035477/download (accessed 10 July 2018).

65. See Hans Robert Jauss, 1982, *Toward an Aesthetic of Reception*, trans. Timothy Bahti, intro. Paul De Man, Minneapolis: University of Minnesota Press.

Chapter 2

1. Margaret Litvin, 2013, 'From Tahrir to "Tahrir": some theatrical impulses toward the Egyptian uprising', *Theater Research International* 38.2.

2. See Jillian Schwedler, 2016, 'Taking time seriously: temporality and the Arab uprisings', *Project on Middle East Political Science*, 3–4 May, https://pomeps.org/2016/06/10/taking-time-seriously-temporality-and-the-arab-uprisings/#_ftn7 (accessed 1 February 2017).

3. See Ahdaf Souheif, 2011, 'Forward', in Nadia Idle and Alex Nunes (eds), *Tweets from Tahrir: Egypt's Revolution as it Unfolded from the People who Made It*, Doha: Bloomsbury, p. 9.

4. Benedict Anderson, 1983, *Imagined Communities*, London: Verso, p. 35.

5. Ibid. p. 35.

6. Martin Heidegger, 1977, *Basic Writings*, New York: Harper and Row, p. 157.

7. Raymond Williams, 1976 [1962], *Communications*, New York: Penguin Books.

8. Habermas points to the case of France, where the "'free communication of ideas and opinions" as it was first articulated in the Constitution of 1791 would later become tied to the "proviso of responsibility for the misuse of this liberty in the cases determined by the law". It was only with the July Revolution of 1830 that the state "gave back to the press and the parties . . . the latitude guaranteed by the revolutionary rights of man".' Jürgen Habermas,

1991 [1962], *The Structural Transformation of the Public Sphere*, Cambridge, MA: MIT Press, pp. 70–1.

9. See, for example, Naila Hamdy, 2016, 'The culture of Arab journalism', in Mohamed Zayani and Suzi Mirgani (eds), *Bullets and Bulletins: Media and Politics in the Wake of the Arab Uprisings*, Oxford: Oxford University Press, p. 72; and Ami Ayalon, 2016, *The Arabic Print Revolution*, Cambridge: Cambridge University Press, pp. 29–30. For a broad survey of newspapers and their role in fostering national imaginary, see Elisabeth Kendall, 2010, *Literature, Journalism and the Avant-garde: Intersection in Egypt*, London: Routledge.

10. Hamdy, 2016, p. 77.

11. The influence of the broadcast media giant *al-Jazeera* on the formation of what Marc Lynch described as an 'Arab Public Sphere' (2006) has been discussed at great length. Noha Mellor offers a particularly good rebuttal of the notion in concert with a renewed call for greater scholarly attention to the role of local media in assessing the formation or deformation of an 'Arab public sphere'. See Mellor, 2007, *Modern Arab Journalism*, Edinburgh: Edinburgh University Press, pp. 73–97.

12. See Alexandra Sandels, 2008, 'A dark year for press freedom in Egypt', *Menassat*, 9 January, http://www.menassat.com/?q=en/news-articles/2641-dark-year-press-freedom-egypt (accessed 2 January 2018).

13. The Muslim Brotherhood had been publishing clandestine print magazines and journals for decades. Noha Mellor traces the transnational publication history of *al-Da'wa* the group's most well-known venture in her 2018 book *Voice of the Muslim Brotherhood: Da'wa, Discourse, and Political Communication*, London: Routledge. Scores of Muslim Brotherhood publications, including new newspapers like *Al-Hurriya wa al-adala*, began appearing in the autumn of 2011 (Mellor, 2018, p. 195).

14. Osama Saraya, 2011, 'Tanzim irhabi min 19 intihariya li-tafjir dur al-'ibada', *Al-Ahram* (Cairo), 25 January, p. 1.

15. Al-Shorouk, 2011 'Youm al-ghadab', *Al-Shorouk* (Cairo), 25 January, p. 1.

16. In looking at Twitter, Philip N. Howard and Muzammil M. Hussain show the number of tweets associated 'most prominently with political uprisings' peaked in Egypt at about 3,400, two weeks after Mubarak's resignation on 11 February. Two weeks prior to his resignation, 34 per cent of those tweets emanated from outside Egypt. See Phillip Howard and Muzammil M. Hussain, 2013, *Democracy's Fourth Wave: Digital Media and the Arab Spring*, Oxford: Oxford University Press, p. 54.

17. The newspaper *Al-Messa'* used the phrase 'Internet youth' to describe their 25 January cover. See N. Hamdy and E. H. Gomaa, 2012, 'Framing the Egyptian uprising in Arabic language newspapers and social media', *Journal of Communication* 62.2, 198.

18. Muhammad Abd al-Hadi, 2011, 'Ihtijajat wa idtirabat was'a fi Libnan', *Al-Ahram* (Cairo), 26 January, p. 1.

19. Al-Ahram, 2011, 'Al-alaf yasharikun fi mudhahirat selmiya bi-al-Qahira wa al-muhafizat', *Al-Ahram* (Cairo), 26 January, p. 1.

20. Al-Shorouk, 2011, 'Burkan al-ghadab yajtah shawari' al-Qahira', *Al-Shorouk* (Cairo), 26 January, p. 1.

21. Al-Shorouk, 2011, '''Unf 'ashwa'i wa qaswa amniya . . .', *Al-Shorouk* (Cairo), 27 January, p. 1.

22. Ibid. p. 1.

23. Al-Ahram, 2011, 'Wafaa 4 wa isaba 118 . . .', *Al-Ahram* (Cairo), 27 January, p. 1.

24. Ibid. p. 1.

25. Tim Ross, Mathew Moore and Steven Swinford, 2011, 'Egypt protests: America's secret backing for rebel leaders behind uprising', *The Daily Telegraph*, 28 January, http://www.telegraph.co.uk/news/worldnews/africaandindian ocean/egypt/8289686/Egypt-protests-Americas-secret-backing-for-rebel-lea ders-behind-uprising.html (accessed 11 October 2017).

26. Ibid.

27. Margaret Scobey, personal interview with the author, 1 November 2016.

28. Tim Ross [*The Telegraph*], personal interview with the author, 1 December 2016.

29. Sissela Bok, 1989, *Lying: Moral Choice in Public and Private Life*, New York: Vintage, p. 283.

30. Ibid. p. 283.

31. Eric Trager, 2011, 'After Tunisia, is Egypt next?', *The Atlantic*, 17 January, http://www.theatlantic.com/international/archive/2011/01/after-tunisia-is-egypt-next/69656/ (accessed 19 October 2017).

32. Alexis Madrigal, 2011, 'Egyptian activists action plan: translated', *The Atlantic*, 27 January, https://www.theatlantic.com/international/archive/2011/01/egyptian-activists-action-plan-translated/70388/ (accessed 15 October 2017).

33. In his memoir, Ghonim writes that 'Ahmed Saleh' had been charged with maintaining the 'Kullena Khaled Said' page, which he did, presumably, even

after being detained on 28 January. Whether the person revealed by WikiLeaks and reported to have been arrested by *The Telegraph* indeed returned to operate the page as early as 1 February, according to Ghonim, or 'Ahmed Saleh' had become a kind of placeholder for anonymity remains unclear. See Wael Ghonim, 2013, *Revolution 2.0: the Power of the People is Greater Than the People in Power: a Memoir*, Boston: Mariner Books/Houghton Mifflin Harcourt, p. 41.

34. Ibid. p. 41.

35. Al-Ahram, 2011, 'Iqalat al-hukuma', *Al-Ahram* (Cairo), 29 January, p. 1.

36. Al-Ahram, 2011, 'Burqiyat diblumasiya hasal 'alayha WikiLeaks', *Al-Ahram* (Cairo), 29 January, p. 8.

37. Hani Shukrallah, 2012, 'Covering the Arab Spring: myths, lies, and truths', Issam Ferres Institute for Public Policy and International Affairs, The American University in Beirut, 21 May, https://www.youtube.com/watch?v=Wu3nwHhAI2U (accessed 1 September 2016).

38. Hamdy and Gomaa, 2012, p. 202.

39. Al-Ahram, 2011, 'Hazr al-tajawwul fi al-Qahira', *Al-Ahram* (Cairo), 29 January, p. 1.

40. Al-Ahram, 2011, 'Rashid: la narghabu fi ittifaq tijara hurra ma'a Amrika', *Al-Ahram* (Cairo), 26 January, p. 1.

41. See 'Liqa' shabab Misr ma'a Jamal Mubarak wa Rashid Mohamed Rashid', YouTube, 26 October 2009, https://www.youtube.com/watch?v=cGCZqctInrc (accessed 20 October 2017).

42. Al-Shorouk, 2011, 'Heikal yatakallam . . .', *Al-Shorouk* (Cairo), 29 January, p. 1.

43. Magdi al-Galad, 2011, 'Mohamed Hassanein Heikal fi hiwar khas', *Al-Masri al-Youm* (Cairo), 1 February, p. 1.

44. Amr El-Shobaki, 2016, 'Madha tabaqqa min thawrat yanayir?!', *Majallat al-dimuqratiyya al-Ahram* 61, 27.

45. The rhetorical motif of 'the people' as it was used in slogans and songs during the eighteen days has been a topic of considerable scholarly attention. See, for example, Elliot Cola, 2012, 'The people want', *The Middle East Research and Information Project (MERIP)*, 42, https://www.merip.org/mer/mer263; Hanan Sabea, 2014, '"I dreamed of being a people": Egypt's revolution, the people and critical imagination', in P. Werbner, M. Webb and K. Spellman-Poots (eds), *The Political Aesthetics of Global Protest: the Arab Spring and Beyond*, Edinburgh: Edinburgh University Press, pp. 67–92; and Noha Mellor, 2016,

The Egyptian Dream: Egyptian National Identity and Uprisings, Edinburgh: Edinburgh University Press, p. 93.

46. Al-Ahram, 2011, 'Ijtima' l-il-ra'is Mubarak ma'a al-qiyadat al-'askariya', *Al-Ahram* (Cairo), 31 January, p. 1.

47. Al-Shorouk, 2011, 'Tard 'Izz min al-watani wa anba' 'an hurub 'Alaa' wa Gamal Mubarak ila London', *Al-Shorouk* (Cairo), 30 January, p. 1.

48. Al-Ahram, 2011, 'Hukuma jadida bi-la 'rijal a'mal', *Al-Ahram* (Cairo), 1 February, p. 1.

49. Al-Ahram, 2011, 'Al-malayin yakharajun', *Al-Ahram* (Cairo), 3 February, p. 1.

50. Al-Jumhuriya, 2011, 'Makasib al-thawra al-sha'biya', *Al-Jumhuriya* (Cairo), 4 February, p. 1.

51. Ibid. p. 1.

52. Al-Ahram, 2011, 'Suleiman: 'al-huwwar' aw 'al-inqilab', *Al-Ahram* (Cairo), 9 February, p. 1.

53. Mohammed Abd al-Rauf, 2011, 'Al-i'lam al-hukumi al-Misri yasara' fi-l-lihaq bi qitar al-thawra', *Al-Sharq al-awsat*, 13 February, http://archive.aawsat.com/details.asp?section=4&article=607997&issueno=11765#.V62PPqJp5WU (accessed 20 October 2017).

54. See Nathaniel Greenberg, 2011, 'Chaos comes to Cairo: neighbors unite to keep the peace', *The Seattle Times*, 31 January, https://www.seattletimes.com/nation-world/chaos-comes-to-cairo-neighbors-unite-to-keep-the-peace-uw-common-language-project/ (accessed 1 May 2018).

55. Al-Masri al-Youm, 2011, 'Al-Nida' al-akhir . . .', *Al-Masri al-Youm* (Cairo), 29 January, p. 1.

56. On 28 January, *Al-Masri al-Youm* featured the former head of the International Atomic Energy Agency on its cover with the dramatic headline 'ElBaradei arrives in Cairo demands complete and immediate change'. The reasons for his disappearance from the news are uncertain. However, at least three major factors appear to have contributed. Firstly, beyond a subset of well-educated, mainly young Egyptians, ElBaradei had never received significant popular support in Egypt. In his memoir, Hamdi Qandil, one of the country's oldest liberal denizens, describes his initial surprise at ElBaradei's desire to run for the highest office. He quotes Jurj Isshaq, a founding member of the Kefaya movement as saying in response to ElBaradei's appearance in Tahrir: 'we don't want to know anything about him' (see Hamdi Qandil, 2014, *'Ishtu marratayn*, Cairo: Dar al-Shorouk, p. 512). Secondly, associates close to ElBaradei's campaign described for me in confidence that he began petitioning the media

at this time to keep himself out of the news. Thirdly, the damning memos released by WikiLeaks, while primarily serving to undercut the April 6 group, implicated by extension everyone associated with the January 25 movement. Undoubtedly, ElBaradei was aware of this and hence, we might assume, his sudden reluctance to receive media attention.

57. Al-Shorouk, 2011, 'Masirat l-mi'at al-alalaf min al-mutazahirin tawasul al-hitaf 'al-sha'b yurid isqat al-nizam', *Al-Shorouk* (Cairo), 30 January, p. 1.

58. Al-Ahram, 2011, 'Al-sha'b asqata al-nizam', *Al-Ahram* (Cairo), 12 February, p. 1.

59. In his 2008 manifesto *Al-Ayyam al-Akhira* (The Last Days), Abd al-Halim Qandil, a founding member of the Kefaya movement and a prolific writer, described five possible fallouts from the imminent 'bankruptcy' of the Mubarak regime. One posited military rule, another the rise of the Brotherhood. The third option, 'dreaded by all parties', was the rule of succession and the inheritance of power by Gamal Mubarak (Qandil, 2008, *Al-Ayyam al-Akhira*, al-Qahira: Dar al-Thaqafa al-Jadida). 'Where the people saw just a leader' he said, 'I saw an entire regime as weak, as sick. A corrupt family with billionaires around it that relied on a dense security apparatus' (Bidoun, 2011, 'Enough is not enough: Abdel Halim Qandil', *Bidoun #25*, http://bidoun.org/articles/enough-is-not-enough (accessed 1 June 2016)). It was a theme he pursued feverishly over the span of a decade beginning with his first book, *Didd al-ra'is: akhtar hamlat maqalat didd hukm al-'a'ilah* (2005). Developed following his abduction and beating by security forces who had kidnapped him after he published several pieces articulating his 'campaign against the institution of the presidency', Qandil said that his writing was provoked by the succession to power of Bashar al-Assad in Syria, where, following his father's death in 2000, the Parliament had voted to change the mandatory age for the presidency to allow the son to inherit the seat. 'That same year', he explained in a 2011 interview with *Bidoun*, 'Gamal Mubarak was in a similar position in the ruling National Democractic Party in this country' (Bidoun, 2011).

60. Ernesto Laclau, 2005, *On Populist Reason*, New York: Verso, p. 87. The emphasis is Laclau's.

61. Heikal's narrative of 'the first complete revolution in Egyptian history' included specific reference to the Free Officer's Coup of 1952 with the assertion that the 2011 uprising had achieved what Nasser first set in motion. Al-Shorouk, 2011, 'Mohamed Hassenein Heikal yattakallam l-il-Shorouk', *Al-Shorouk* (Cairo), 3 February, p. 1. The phrase: 'the Sense of an Ending' is borrowed from the

eponymous title of Frank Kermode's famous 1967 study of mortality and eternity in Western literature.

62. Al-Shorouk, 2011, 'Al-Sha'b yataqaddam wa Mubarak yabda' al-taraju'', *Al-Shorouk* (Cairo), 30 January, p. 1.

63. Laclau, 2005, p. 83.

64. Al-Shorouk, 2011, 'Muqtarahat 'Omar Suleiman l-ihtiwa' intifadat al-ghadab', *Al-Shorouk* (Cairo), 31 January, p. 1.

65. 'I have a Ph.D. in stubbornness' Mubarak famously said after the Million Man march. The phrase was reported in headlines of *Al-Masri al-Youm*, 2 February 2011.

66. Anderson, 1983, p. 33.

67. Hans Robert Jauss, 1982, *Toward an Aesthetic of Reception*, trans. Timothy Bahti, intro. Paul De Man, Minneapolis: University of Minnesota Press, p. 53.

68. Laclau, 2005, p. 88.

69. Ibid. p. 88.

70. Margaret Scobey noted that she was sceptical about some of the State Department's Diplomacy 2.0 initiatives, in part because they relied on under-experienced operators at the expense of Egypt's more established opposition forces. Scobey, 2016.

Chapter 3

1. Portions of this chapter appeared in the e-zine *Jadaliyya* as 'The rise and fall of Abu 'Iyadh: reported death leaves questions unanswered', 13 July 2015.

2. For a good, brief discussion on the distinction between the 'Salafi-jihadist' tendency, as opposed to the 'Salafi-quietest' or 'scientific' one, see Gilles Paris's interview in *Le Monde* with French researcher Stéphane Lacroix, 2012, 'Le salafisme, c'est le dogme dans toute sa pureté', *Le Monde*, 27 September, http://www.lemonde.fr/culture/article/2012/09/27/le-dogme-dans-toute-sa-purete_1766968_3246.html#yYzm5pyIwbviHdAO.99 (accessed 1 October 2013).

3. Foucault points to the example of Eugène François Vidocq (1775–1857) in discussing the 'utility of crime' to the stability of the First Republic follow-ing the collapse of monarchy in France. See Michel Foucault, 1980, *Power/Knowledge: Selected Interviews and Other Writings, 1972–1977*, New York: Pantheon Books, p. 45.

4. Six thousand Tunisians were believed to have travelled to Iraq and Syria to wage jihad. The number provided by the Soufan Group in 2016 is nearly triple

that of the second largest pool of fighters from Saudi Arabia. See Yaroslav Trofimov, 2016, 'How Tunisia became a top source of Tunisian recruits', *The Wall Street Journal*, 25 February, https://www.wsj.com/articles/how-tunisia-became-a-top-source-of-isis-recruits-1456396203 (accessed 9 March 2018).

5. See M. Khayat, 2012, 'The rise of the Salafi-Jihadi movement in Tunisia – the case of Ansar Al-Shari'a in Tunisia', The Middle East Media Research Institute, 20 March, http://www.memrijttm.org/the-rise-of-the-salafi-jihadi-movement-in-tunisia--the-case-of-ansar-al-sharia-in-tunisia.html (accessed 1 April 2012).

6. *Tunisie numerique* was one of several online Francophone websites to report biographical information on Ayadh without any source. Most of the information, however, paralleled what Ayadh himself disclosed in the un-translated interview with 'Walid Yusuf' published on *Muslim.org*. C.f. 'Abu Iyadh: L'illustre inconnu', *Tunisie numerique*, 18 September 2012, http://www.tunisienumerique.com/abou-iyadh-lillustre-inconnu/144762 (accessed 18 September 2012).

7. Walid Yusuf, 2012, 'al-Liqa' al-sahafi ma'a Abi 'Ayadh al-Tunisi', *Muslim.org*, 1 January, http://www.muslm.org/vb/archive/index.php/t-465707.html (accessed 1 October 2012).

8. Al-Karama, 2007, 'Tunisia risk of violation of Sayfallah Ben Hassine's right to life', *Al-Karama for Human Rights*, 17 July, https://www.alkarama.org/en/articles/tunisia-risk-violation-sayfallah-ben-hassines-right-life (accessed 19 January 2018).

9. United Nations Security Council, 2017, *Consolidated United Nations Security Council Sanctions List*, https://www.un.org/sc/suborg/en/sanctions/un-sc-consolidated-list (accessed 26 February 2018).

10. Ould Ellil fell under heavy scrutiny in the run-up to the 26 October 2014 elections, when police raided a house adjacent to the mosque where it was believed Abu Ayadh adjourned the first meetings of Ansar al-Sharia. Half a dozen people were killed in the shootout that followed. See Frida Dahmani, 2014, 'Tunisie: Tension sécuritaire à trois jours des legislatives', *Jeune Afrique*, 23 October, http://www.jeuneafrique.com/Article/ARTJAWEB20141023143123/tunisie-terrorisme-legislatives-tunisiennes-terrorisme-tunisie-tension-securitaire-a-trois-jours-des-legislatives.html.

11. Fatima Jilasi, 2014. 'Takshifu asrar wa ahdaf ta'sis Ansar al-Shari'a', *Al-Sabah*, 6 March, http://www.assabah.com.tn/article/80924/%D8%A7%D9%84%D8%B5%D8%A8%D8%A7%D8%AD-%D8%AA%D9%83%D8%B4%D9%81-%D8%A3%D8%B3%D8%B1%D8%A7%D8%B1-%D9%88%

D8%A7%D9%87%D8%AF%D8%A7%D9%81-%D8%AA%D8%A3%
D8%B3%D9%8A%D8%B3-%D8%A7%D9%86%D8%B5%D8%A7%
D8%B1-%D8%A7%D9%84%D8%B4%D8%B1%D9%8A%D8%B9%
D8%A9.

12. Mohamed Bughlab, 2012, 'Al-Shaykh 'Khamis al-Majri' fi hiwar khas ma'a
al-Tunisiya', *Al-Tunisiya*, 18 April, http://www.attounissia.com.tn/details_arti
cle.php?t=37&a=56222 (accessed 1 October 2012).

13. The chant, '*Khaybr, khaybr ya yahud*' ('Khaybr, Khaybr Oh Ye Jew') is in reference to the seventh century battle between Jews and Muslims in the northwest-
ern quarter of the Arabian Peninsula. '*Obama, Obama, koluna Osama*' became
a common refrain. It means 'Obama we are all Osama'. This quote, as with the
image in Figure 3.1, was taken from video recordings posted by the group on
YouTube. They are no longer available for public view.

14. A more immediate impetus for the attack on the art exhibition, 'Le printemps
d'art' – which featured, among other things, an installation of two women in
hijab surrounded by stones, another with the word of God formed by ants
crawling from the nose of a young boy, and a painting of a semi-nude woman
against the backdrop of bearded men – was likely a call earlier that month by
a Salafi shaykh in Tunisia, Hussayn Obaydi. See Haim Malka, 2015, 'Tunisia:
confronting extremism', in J. B Alterman (ed.), *Religious Radicalism after the
Arab Uprisings*, Lanham: Rowman and Littlefield, p. 115.

15. See Ettounsiya TV, 2014, 'Ali Laârayedh affirme avoir décidé de ne pas arrêter
Abou Iyadh', YouTube, 1 June, https://www.youtube.com/watch?v=-JBzLX
DIv9s (accessed 17 January 2018).

16. C.f. Directinfo, 2012, 'Arrestation d'un Tunisien par les garde-frontières
algériens pour contrebande', *Directinfo.com*, 25 October, http://directinfo.
webmanagercenter.com/2012/10/25/arrestation-dun-tunisien-par-les-garde-
frontieres-algeriens-pour-contrebande/ (accessed 25 October 2012).

17. Mosaique FM, 2014, 'Abou Iyadh se cachait à proximité du ministère de
l'Intérieur', *Mosaique FM*, 3 March, http://archivev2.mosaiquefm.net/fr/
index/a/ActuDetail/Element/35191-abou-iyadh-se-cachait-a-proximite-du-
ministere-de-l-interieur (accessed 17 January 2018).

18. See Al-Tunisiya, 2014, 'Abu Loqman yakhlufu Abu 'Ayadh 'ala ra's Ansar
al-Shari'a fi Tunis', *Al-Tunisiya*, 27 February, http://www.attounissia.com.tn/
details_article.php?t=42&a=114779 (accessed 27 February 2014).

19. See Benjamin Roger, 2013, 'Tunisie: sur la trace des jihadistes du mont
Chaambi', *Jeune Afrique*, 5 July, http://www.jeuneafrique.com/Article/

ARTJAWEB201305071618/alg-rie-tunisie-terrorisme-al-qaeda-terrorisme-tunisie-sur-la-trace-des-jihadistes-du-mont-chaambi.html (accessed 5 July 2013).

20. Sabah al-Shabbi, 2014, 'Abu 'Ayadh', *Al-Sabah*, 18 October, http://www.assa bahnews.tn/article/92944/%D8%B4%D9%82%D9%8A%D9%82%D9% 87-%D9%8A%D9%81%D8%AC%D9%91%D8%B1-%D9%85%D9% 81%D8%A7%D8%AC%D8%A3%D8%A9-%D9%85%D9%86-%D8% A7%D9%84%D9%88%D8%B2%D9%86-%D8%A7%D9%84%D8% AB%D9%82%D9%8A%D9%84-%D8%A3%D8%A8%D9%88-%D8% B9%D9%8A%D8%A7%D8%B6-%D9%8A%D8%B1%D8%B3%D9% 84-%D8%A7%D9%84%D8%A3%D9%85%D9%88%D8%A7%D9% 84-%D9%85%D9%86-%D9%84%D9%8A%D8%A8%D9%8A%D8% A7-%D9%81%D9%8A-%D8%B9%D9%84%D8%A8-%D8%A7%D9% 84%D8%B4%D9%88%D9%83%D9%88%D9%84%D8%A7%D8%B7% D8%A9 (accessed 26 January 2015).

21. Mosaique FM, 2014, 'Bin Jidu: Abu 'Ayadh fi Derna', *Mosaique FM*, 29 November, https://www.facebook.com/mosaiquefm/posts/%D8%A8%D9% 86-%D8%AC%D8%AF%D9%88-:%D8%A3%D8%A8%D9%88-% D8%B9%D9%8A%D8%A7%D8%B6-%D9%81%D9%8A/10153022451 153646/ (accessed 10 December 2018).

22. In addition to the assassination of the deputy minister, news of a 'special forces officer' killed in Derna surfaced in a *Voice of America* report concerning an increase in tribal violence in the south of the country. The article published by the official US government organ provides no additional detail on the unnamed 'officer'. See Voice of America, 2014, 'Tribal leaders unveil "Save Libya" plan', *VOA*, 13 January, https://www.voanews.com/a/tribal-leaders-unveil-save-libya-plan/1829342.html (accessed 22 January 2018).

23. Le Temps, 2014, 'La HAICA face aux desperados', *Le Temps*, 15 October, http://www.letemps.com.tn/article/86856/la-haica-face-aux-desperados (accessed 15 October 2014).

24. The Director General of Security in Tunisia, Nabil Obayd, testified against Laaraydh in January 2015, that he had been given orders from the Ministry of the Interior to help Ayadh escape from the Mosque. Laaraydh denied the accusation, reiterating claims made at the time that Obayd, as head of security, had made the decision not to pursue Ayadh inside the mosque because of congestion in the downtown vicinity. But claims that the Ennahdha Interior Minister was aware of, if not complicit, in the movement of Abu Ayadh continued to

gain momentum after his disappearance. In March 2014, an official inquiry into the assassination of the labour leaders found that in late 2012 Ayadh had relocated to a house in central Tunis, in direct proximity to the Ministry of the Interior. See Mosaique FM, 3 March 2014.

25. Ghazi, 2014, 'Terrorisme: L'frère de Abou Iyadh se mettre-ti-il à table', *Le Temps*, 12 October, http://www.letemps.com.tn/article/86778/terrorisme-le-fr%C3%A8re-de-abou-iyadh-se-mettra-t-il-%C3%A0-table (accessed 24 February 2018).

26. See Frida Dahmani, 2012, 'Tunisie: les salafistes, ces très inquiétants fous de Dieu', *Jeune Afrique*, 20 June, http://www.jeuneafrique.com/141108/politique/tunisie-les-salafistes-ces-tr-s-inqui-tants-fous-de-dieu/ (accessed 16 January 2018).

27. Rached Ghannouchi and Olivier Ravanello, 2015, *Entretiens d'Olivier Ravanello avec Rached Ghannouchi*, Paris: Plon, p. 19.

28. See Al-Arabiya, 2013, 'Abu 'Ayadh al-Tunisi', *al-Arabiya*, 12 February, https://www.alarabiya.net/ar/arab-and-world/2013/02/12/%D8%A3%D8%A8%D9%88-%D8%B9%D9%8A%D8%A7%D8%B6-%D8%A7%D9%84%D8%AA%D9%88%D9%86%D8%B3%D9%8A-%D8%A7%D8%AE%D8%AA%D8%B7%D8%A7%D9%81-%D8%B1%D9%87%D8%A7%D8%A6%D9%86-%D8%A7%D9%84%D8%AC%D8%B2%D8%A7%D8%A6%D8%B1-%D8%A5%D8%B1%D9%87%D8%A7%D8%A8-%D9%85%D8%AD%D9%85%D9%88%D8%AF--1418.html (accessed 28 February 2018).

29. Material captured from the Ansar al-Sharia Facebook page is no longer publicly available. As with the material from AQIM, research on Ansar al-Sharia was harvested in 2012 as part of a now-concluded study with the Center for Strategic Communication at Arizona State University.

30. Aaron Zeilin, 2012, 'Know your Ansar al-Sharia', *Foreign Policy*, 21 September, http://foreignpolicy.com/2012/09/21/know-your-ansar-al-sharia/ (accessed 17 January 2018).

31. Shaykh Abu-Muslim al-Jazairi, 2011, 'Fatwa supports "Campaign to defend the Islamic identity of Tunisia"', *Al-Andalus Establishment for Media Production*, http://www.opensource.gov (accessed 18 May 2013).

32. Article 1 of both the 1959 and 2015 Tunisian Constitution reads: 'Tunisia is a free, independent and sovereign state. Its religion is Islam, its language is Arabic and its type of government is the Republic.'

33. Al-Jazairi, 2011.

34. Video of Abu Ayadh's 2012 speech in Kairouan is no longer readily available on YouTube. The quotation stems from the author's notes in 2012.

35. See Anne Wolf, 2017, *Political Islam in Tunisia*, Oxford: Oxford University Press, p. 144.

36. Salah Abu Muhammad, 2011, 'AQLIM says *Al-Hayat* interview with spokesman "fake"', *Al-Andalus Establishment for Media Production*, http://www.opensource.gov (accessed 20 May 2013). Among the 'falsities' the speaker notes are claims that AQIM had: 'established an Islamic emirate in the eastern part of Libya'; referred to the Libyan National Interim Council as 'the interim infidel council'; had 'gained possession over surface-to-air missiles'; and negotiated with the French 'to receive ransom against releasing their captives'.

37. Ibid.

38. Abu Abd al-Rahman, 2010 'Letter to Salah Abu Muhammad', *Dni.gov*, https://www.dni.gov/index.php (accessed 18 January 2018). The italics are mine.

39. Shaykh Abu-Hayyan Asim, 2010, 'Audio message from AQLIM Shar'iah [sic] Commission Member' [Open letter to the people of the frontiers and their supporters by Shaykh Abu Hayyan Asim (May God protect him)], *Al-Andalus Establishment for Media Production*, http://www.opensource.gov (accessed 10 May 2013).

40. Sissela Bok, 1989, *Lying: Moral Choice in Public and Private Life*, New York: Vintage, p. 5.

41. Ibid. p. 5.

42. Philippe-Joseph Salazar, 2017, *Words are Weapons: Inside ISIS's Rhetoric of Terror*, New Haven: Yale University Press, p. 154.

43. Jeffry R. Halverson, H. L. Goodall and Steve Corman identified the discursive trope of 'Satan's handiwork' and the act of deception as a 'master narrative' of Islamic extremism. See Halverson *et al.*, 2011, *Master Narratives of Islamic Extremism*, New York: Palgrave, pp. 125–37.

44. Bok, 1989, p. 7.

45. Asim, 2010.

46. Fabio Merone, 2013, 'Salafism in Tunisia: an interview with a member of Ansar al-Sharia', *Jadaliyya*, 11 April, http://www.jadaliyya.com/pages/index/11166/salafism-in-tunisia_an-interview-with-a-member-of (accessed 24 February 2018).

47. The phrase 'Old Man of the Mountain' was used by Richard I of England, Omar Khayyam and others in reference to Hassan al-Sabbah (d. 1124), leader

of the eleventh-century *Shiaa* sect the *Hashashin* or 'The Assassins'. In his monumental study of the subject, *The Order of the Assassins* (1955), Marshall G. S. Hodgson noted that the legend entered into 'Western lore' with Edward FitzGerald's translation of Omar Khayyam's poetry. It would become so embedded in the Western imagination by the twelfth century that Richard I of England (d. 1199) 'was accused of imitating the Old Man of the Mountain in training murderers; and found it necessary to have a group of men in England plead guilty to such operations against himself'. The Hashashin assassinated numerous political figures, including Nizam al-Mulk (d. 1092), the vizier of the Seljuk Empire and two caliphs: Mustarshid (d. 1135) and Rashid (d. 1138). See Marshall G. S. Hodgson, 1980, *The Order of Assassins*, New York: AMS Press, p. 138.

48. See Jacques Derrida, 1999, *The Gift of Death*, Chicago: University of Chicago Press, p. 65.

49. Ibid. p. 65.

50. Al-Arabiya, 12 February 2013.

51. Limited information concerning the Guantánamo prisoners as well as the countries they were transferred to is available on *The New York Times* website, https://www.nytimes.com/interactive/projects/guantanamo/transfer-coun tries/kazakhstan (accessed 23 January 2018).

52. RT, 2015, 'Al-Jaysh al-Suri yath'ar l-il-Kisabsa', *RT*, 10 March, https://arabic. rt.com/news/795736-%D8%AC%D9%8A%D8%B4-%D8%B3%D9%88% D8%B1%D9%8A-%D8%A3%D8%A8%D9%88-%D8%A8%D9%84% D8%A7%D9%84-%D8%A7%D9%84%D8%AA%D9%88%D9%86% D8%B3%D9%8A-%D8%BA%D8%A7%D8%B1%D8%A9-%D8%AC% D9%88%D9%8A%D8%A9/ (accessed 24 February 2018).

53. Christopher Hope, Robert Winnett, Holly Watt and Heidi Blake, 2011, 'WikiLeaks: Guantanamo Bay terrorist secrets revealed', *The Telegraph*, 25 April, http://www.telegraph.co.uk/news/worldnews/wikileaks/8471907/Wiki Leaks-Guantanamo-Bay-terrorist-secrets-revealed.html (accessed 22 January 2018).

54. Trump repeated the claim on multiple occasions during the 2016 presidential election. Michael McFaul, a former Ambassador to Russia, among others, pointed out that Russian state media services had been promulgating the narrative of US complicity with ISIS for over a year. Breitbart News, whose former Editor-in-Chief Steve Banon would become an early cabinet appointee of President Trump, was among the principal US outlets promoting the false

narrative. See Erik Prince, 2016, 'Obama and Clinton are complicit in creating ISIS', *Breitbart.com*, 16 June, http://www.breitbart.com/national-security/2016/06/16/erik-prince-obama-clinton-complicit-creating-isis/ (accessed 25 January 2018).

55. Hans-Georg Gadamer, 1975, *Truth and Method*, New York: Continuum, p. 299.

56. Gabriel Levinas, 1998, *La ley bajo los escombros: AMIA: Lo que no se hizo*, Buenos Aires: Editorial Sudamericana, p. 87. Quoted in N. Greenberg, 2010, 'War in pieces: AMIA and the triple frontier in Argentine and American discourse on terrorism', *A Contracorriente* 8.1.

57. See Aziz Krichen, 2016, *La Promesse du printemps*, Paris: Script. An excerpt of the work is available at https://nawaat.org/portail/2016/04/02/bonnes-feuilles-la-promesse-du-printemps-ce-aziz-krichen/.

58. See my discussion on 'Lab 404' and the hacking of the regime's censorship machine in Chapter 6.

59. See D. G. Tor, 2015, 'God's cleric: Al-Fuḍayl b. 'Iyāḍ and the transition from caliphal to prophetic Sunna', in Behnam Sadeghi, Asad Q. Ahmed, Adam J. Silverstein and Robert G. Hoyland (eds), *Islamic Cultures, Islamic Contexts: Essays in Honor of Professor Patricia Crone*, Leiden: Brill.

60. Plato, 1992, *The Republic*, trans. G. M. A. Grube and C. D. C. Reeve, Indianapolis: Hackett Publishing Co., p. 232.

61. See Christopher Dickey, 2008, 'Using comics to turn off terror', *Newsweek*, 17 April, http://www.newsweek.com/using-comics-turn-terror-86433 (accessed 1 May 2013).

Chapter 4

1. Bassam Haddad, 2016, 'The debate over Syria has reached a dead end', *The Nation*, 18 October, https://www.thenation.com/article/the-debate-over-syria-has-reached-a-dead-end/ (accessed 21 July 2017).

2. Alaa Abd El Fattah, 2016, ' "I was terribly wrong" – writers look back at the Arab Spring five years on', *The Guardian*, 3 January, https://www.theguardian.com/books/2016/jan/23/arab-spring-five-years-on-writers-look-back (accessed 1 February 2018).

3. Human Rights Watch, 2014, 'All according to plan: the Rab'a [sic] massacre and mass killings of protesters in Egypt', *Human Rights Watch*, 12 August, https://www.hrw.org/report/2014/08/12/all-according-plan/raba-massacre-and-mass-killings-protesters-egypt (accessed 6 June 2017).

4. Jean-Pierre Filiu, 2015, *From Deep State to Islamic State*, New York: Oxford University Press, p. 175.

5. See David D. Kirkpatrick, 2014, 'Prolonged fight feared in Egypt after bombings', *The New York Times*, 24 January, https://www.nytimes.com/2014/01/25/world/middleeast/fatal-bomb-attacks-in-egypt.html (accessed 24 February 2018).

6. See Robert Mackey and Liam Stack, 2012, 'Obscure film mocking Muslim prophet sparks anti-U.S. protests in Egypt and Libya', *The New York Times*, 11 September, https://thelede.blogs.nytimes.com/2012/09/11/obscure-film-mocking-muslim-prophet-sparks-anti-u-s-protests-in-egypt-and-libya/?_r=0 (accessed 27 October 2017).

7. Ibid.

8. Muhammad Abd al-Shokur, 2012, 'Video: Al-Nas tadhi' laqatat min al-film al-musi' l-il-rasul', *Al-Wafd*, 9 September, https://alwafd.org/%D8%AF%D9%86%D9%8A%D8%A7-%D9%88%D8%AF%D9%8A%D9%86/262446-%D9%81%D9%8A%D8%AF%D9%8A%D9%88-%D8%A7%D9%84%D9%86%D8%A7%D8%B3-%D8%AA%D8%B0%D9%8A%D8%B9-%D9%84%D9%82%D8%B7%D8%A7%D8%AA-%D9%85%D9%86-%D8%A7%D9%84%D9%81%D9%8A%D9%84%D9%85-%D8%A7%D9%84%D9%85%D8%B3%D9%89%D8%A1-%D9%84%D9%84%D8%B1%D8%B3%D9%88%D9%84 (accessed 30 October 2017).

9. Walid Abd Al-Rahman, 2012, 'Ramzi: Akthar min 100 alf Qibti Misri taqdimu bi-il-talbat hijra', *Al-Arabiya*, 10 September, https://www.alarabiya.net/articles/2012/09/10/237094.html (accessed 30 October 2017).

10. See Matt Bradley, 2012, 'Missions stormed in Libya, Egypt', *WSJ*, 12 September, https://www.wsj.com/articles/SB10000872396390444401750457645681057498266 (accessed 30 October 2017).

11. Amru Ahmed al-Ansari, 2014, *Jumhuriyat al-Ultras: 7 sanawat ashghl shaqqa*, Cairo: Nahdha Misr lil-Nashr, p. 81.

12. See Gleb Bryanski, 2012, 'Youtube under threat in Russia over prophet film', *Reuters*, 18 September, http://www.reuters.com/article/us-protest-russia/youtube-under-threat-in-russia-over-prophet-film-idUSBRE88H16L20120918 (accessed 1 November 2017).

13. Daniel Greenfield, 2012, 'Christopher Stevens feeds the crocodile', *Front Page Magazine*, 12 September, http://www.frontpagemag.com/fpm/144003/christopher-stevens-feeds-crocodile-daniel-greenfield (accessed 31 October 2017).

14. The Telegraph, 2011, 'Die hard in Derna', *The Telegraph*, 31 January, http://www.telegraph.co.uk/news/wikileaks-files/libya-wikileaks/8294818/DIE-HARD-IN-DERNA.html (accessed 1 November 2017).

15. Greenfield, 2012.

16. WikiLeaks and Russian state media would continue to feed this narrative to the American public. In one particularly iconic instance in 2016, WikiLeaks disseminated a State Department Cable concerning a 2012 meeting between Secretary Clinton and then President Mohamed Morsi. The document, which appeared to have been sent directly to *Breitbart News* and *The Washington Free Beacon*, appeared with the headlines 'Bill Clinton boasts of Hilary's working relationship with President Morsi' and 'Clinton backed Egypt's Muslim Brotherhood regime'. Although, according to *The New York Times*, the so called 'secret documents' contained little more than what had already been reported, their release in the midst of the US presidential election appeared calibrated to drive the narrative that Clinton, like Obama, was engaged in a clandestine campaign to Islamise the country.

17. See ONTV, 2012, ''Alaqa Amrika bi-l-Ikhwan al-Muslimin', *ONTV*, 31 December, https://www.youtube.com/watch?v=QCKwsa8KuEk (accessed 2 November 2017).

18. Ibid.

19. See Amr Ammar, 2014, *Al-Ihtilal al-madani*, Cairo: Tawzi al-Majmu'a al-dawliya l-il-nashr wa al-tawzi', pp. 307–23. As late as 2017, Ammar's book was being broadly promoted in Egypt. As reported by *Youm 7*, in January 2017, the book was discounted at the annual Cairo Book Fair as part of an educational effort on the part of the publisher. Youm 7, 2017, 'Bi-ma'radh al-kitab . . .', *Youm 7*, 24 January, http://www.youm7.com/story/2017/1/24/%D8%A8%D9%85%D8%B9%D8%B1%D8%B6-%D8%A7%D9%84%D9%83%D8%AA%D8%A7%D8%A8-%D8%B3%D9%84%D8%B3%D9%84%D8%A9-%D8%A7%D9%84%D8%A7%D8%AD%D8%AA%D9%84%D8%A7%D9%84-%D8%A7%D9%84%D9%85%D8%AF%D9%86%D9%89-%D9%80-%D8%B9%D9%85%D8%B1%D9%88-%D8%B9%D9%85%D8%A7%D8%B1-%D8%B9%D9%86-%D8%AF%D8%A7%D8%B1/3070350 (accessed 8 January 2018).

20. Michele Dunne, personal interview with the author, 14 May 2018.

21. RT, 2013, 'Isra'il wa Suriya wa Iran: milaffat Kerry al-hayawiya l-il-Ikhwan Misr', *RT*, 4 March, https://arabic.rt.com/news/609314-%D8%A5%D8%B3%D8%B1%D8%A7%D8%A6%D9%8A%D9%84_%D9%88%D8%

B3%D9%88%D8%B1%D9%8A%D8%A7_%D9%88%D8%A5%D9%
8A%D8%B1%D8%A7%D9%86_%D9%85%D9%84%D9%81%D8%
A7%D8%AA_%D9%83%D9%8A%D8%B1%D9%8A_%D8%A7%D9%
84%D8%AD%D9%8A%D9%88%D9%8A%D8%A9_%D9%84%D8%
A5%D8%AE%D9%88%D8%A7%D9%86_%D9%85%D8%B5%D8%
B1/ (accessed 15 November 2017).

22. Peter Pomerantsev, 2014b, 'Russia and the menace of unreality', *The Atlantic*, 9 September, https://www.theatlantic.com/international/archive/2014/09/ russia-putin-revolutionizing-information-warfare/379880/ (accessed 24 February 2018). This begs the question: is counter-factual storytelling effective in disrupting existing narratives (i.e. the official US line that the relationship between Washington and the Brotherhood was pragmatic)? Anecdotal research suggests that indeed it may be. Yanick Bressan and researchers at the French Center for Intelligence Research for example observed in a clinical neorological trial of twenty spectators at a controlled theatrical performance that some 80 per cent revealed neurological signs of '*adhesion*', that is, the 'displacement of perception' from the 'immediately perceptible' (i.e. people on stage) to the fictitiously induced (i.e. actors in a story). See Safouane Hamdi and Yannick Bressan, 2013, 'Du théâtre à l'amphithéâtre: pour une extension du concept d'adhésion à la neuroscience éducationnelle', *Neuroéducation* 2.1.

23. See Engy Abdelkader, 'The anatomy of a terrorist designation: the Muslim Brotherhood and international terrorism', Penn Law Global Affairs Blog, https://www.law.upenn.edu/live/news/6858-the-anatomy-of-a-terrorist-designation-the-muslim/news/international-blog.php (accessed 1 May 2018).

24. Robert Booth, 2010, 'Cables claim al-Jazeera changed coverage to suit Qatari foreign policy', *The Guardian*, 5 December, https://www.theguardian.com/ world/2010/dec/05/wikileaks-cables-al-jazeera-qatari-foreign-policy (accessed 30 May 2018).

25. See, for example, The Economic Times, 2010, 'Qatar uses Al-Jazeera as bargaining chip: WikiLeaks', *The Economic Times*, 6 December, https://economictimes. indiatimes.com/tech/internet/qatar-uses-al-jazeera-as-bargaining-chip-wikile aks/articleshow/7051690.cms (accessed 30 May 2018); Basma Elbaz, 2017, 'Stop victimizing al-Jazeera', *HuffPost*, 29 June, https://www.huffingtonpost. com/entry/stop-victimizing-al-jazeera_us_5954f466e4b0f078efd98794 (accessed 30 May 2018).

26. Adam Taylor, 2016, 'When Trump calls Obama the "founder of ISIS" he sounds like a Middle East conspiracy theorist', *The Washington Post*, 11

August, https://www.washingtonpost.com/news/worldviews/wp/2016/08/11/
when-trump-calls-obama-the-founder-of-isis-he-sounds-like-a-middle-east-
conspiracy-theorist/?utm_term=.0aae9965ce98 (accessed 12 August 2016).

27. See Michael McFaul, 2016, Twitter, 11 August, https://twitter.com/McFaul/
status/763581341984096256?ref_src=twsrc%5Etfw&ref_url=http%3A%
2F%2Fwww.stopthedonaldtrump.com%2Ftag%2Fisis%2F (accessed 25
January 2018).

28. See Nathaniel Greenberg, 2016, 'Exit ISIS, stage left: fighting for laughs
in Mosul and beyond', *Jadaliyya*, 16 April, http://www.jadaliyya.com/
Details/33178/Exit-ISIS,-Stage-Left-Fighting-for-Laughs-in-Mosul-and-
Beyond (accessed 24 February 2018).

29. See Aryn Baker, 2014, 'Why Iran believes the militant group ISIS is an
American plot', *Time*, 19 July, http://time.com/2992269/isis-is-an-american-
plot-says-iran/ (accessed 24 February 2018).

30. Noha Mellor addresses the Brotherhood's communication strategies following
the 2011 uprising, including their rhetorical emphasis on foreign affairs, in
her superb 2018 book: *Voice of the Muslim Brotherhood: Da'wa, Discourse, and
Political Communication*, London: Routledge, pp. 195–6.

31. See Nathaniel Greenberg, 2012, 'The Arab constitutions: chaos and strategy',
COMOPS, 1 December, http://csc.asu.edu/2012/12/01/the-arab-constitu
tions-2012-chaos-and-strategy/ (accessed 24 February 2018).

32. Mustafa Bakri, 2015, *Al-Sisi. al-tariq ila bina' al-dawla*, Cairo: Dar al-Masriya
al-Libnaniya, p. 6.

33. Belal Fadl, 2014, 'The political marshal of Egypt', *Mada Masr*, 2 February,
http://www.madamasr.com/en/2014/02/02/opinion/u/the-political-marshal-
of-egypt/ (accessed 23 May 2017).

34. Ibid.

35. For more on the '*eid wahda*' phenomenon and the march towards Tamarrod,
see Walter Armbrust, 2013, 'The trickster in Egypt's January 25 Revolution',
Comparative Studies in Society and History 55.4, 834–64.

36. See Al-Bernameg, 2014, *YouTube*, https://www.youtube.com/watch?v=1Cu
ZnRVGinw (accessed 24 February 2018).

37. Quoted in Mada Masr, 2014, 'Renowned novelist Aswany quits column, citing
censorship', *Mada Masr*, 24 June, http://www.madamasr.com/en/2014/06/24/
news/u/renowned-novelist-aswany-quits-writing-column-citing-censorship/
(accessed 23 May 2017).

38. Ayman Ramadan, 2017, 'Wa'il al-Ibrashi', *Youm 7*, 20 October, http://www.

youm7.com/story/2014/10/20/%D9%88%D8%A7%D8%A6%D9%84-%
D8%A7%D9%84%D8%A5%D8%A8%D8%B1%D8%A7%D8%B4%
D9%89-%D9%88%D9%82%D9%81-%D8%A7%D9%84%D8%A8%
D8%AB-%D8%B9%D9%86-%D8%A7%D9%84%D8%B9%D8%A7%
D8%B4%D8%B1%D8%A9-%D9%85%D8%B3%D8%A7%D8%A1-%
D9%84%D8%A3%D8%B3%D8%A8%D8%A7%D8%A8-%D8%B3%
D9%8A%D8%A7%D8%B3%D9%8A%D8%A9-%D9%88%D9%84%
D9%8A%D8%B3%D8%AA/1913602 (accessed 23 May 2017).

39. Heba Afify, 2014, 'Egyptian media isn't taking prisoners, State's line is only line', *Mada Masr*, 27 October, http://www.madamasr.com/en/2014/10/27/feature/politics/egyptian-media-isnt-taking-prisoners-states-line-is-only-line/ (accessed 23 May 2017).

40. Mahmoud Mourad, 2015, 'Egyptian poet goes on trial accused of contempt of Islam', *Reuters*, 28 January, http://www.reuters.com/article/us-egypt-courts-poet-idUSKBN0L121M20150128 (accessed 23 May 2017).

41. Samir Farid, 2002, *Tarikh al-raqaba 'ala al-sinima fi Misr*, al-Qahira: al-Maktab al-Misri li-Tawzi' al-Matbu'at, p. 58.

42. Ami Ayalon, 2016, *The Arabic Print Revolution: Cultural Production and Mass Readership*, Cambridge: Cambridge University Press, p. 66.

43. Khalid Kishtainy, 1985, *Arab Political Humor*, London: Quartet Books, p. 85.

44. Ibid. p. 85.

45. Ibid. p. 85.

46. Farid, 2002, p. 59. Also quoted in Nathaniel Greenberg, 2014, *The Aesthetic of Revolution in the Film and Literature of Naguib Mahfouz*, Lanham: Lexington Books, p. 4.

47. Greenberg, 2014, p. 4.

48. Ibid. p. 4.

49. Le Monde, 2011, 'L'Egypt: autre "ennemi d'Internet"', *Le Monde*, 26 January, http://www.lemonde.fr/technologies/article/2011/01/26/l-egypte-autre-ennemi-d-internet_1470630_651865.html (accessed 13 July 2017).

50. Wael Abbas, 2007, 'Help our fight for real democracy', *The Washington Post*, 27 May, http://www.washingtonpost.com/wp-dyn/content/article/2007/05/25/AR2007052502024_pf.html (accessed 13 July 2017).

51. Ibid.

52. Mohamed As'ad, 2016, 'Hay'at jadida l-il-sihafa wa al-i'lam', *Youm 7*, 19 November, http://www.youm7.com/story/2016/11/19/3-%D9%87%D9%

8A%D8%A6%D8%A7%D8%AA-%D8%AC%D8%AF%D9%8A%D8%
AF%D8%A9-%D9%84%D9%84%D8%B5%D8%AD%D8%A7%D9%
81%D8%A9-%D9%88%D8%A7%D9%84%D8%A5%D8%B9%D9%
84%D8%A7%D9%85-%D9%86%D9%86%D8%B4%D8%B1-%D9%
85%D9%84%D8%A7%D9%85%D8%AD-%D8%AA%D8%B4%D9%
83%D9%8A%D9%84-%D8%A7%D9%84%D9%85%D8%AC%D9%
84%D8%B3-%D8%A7%D9%84%D8%A3%D8%B9%D9%84%D9%
89/2974004 (accessed 20 July 2017).

53. Ruth Michaelson, 2017, 'Egypt blocks access to news websites including Al-Jazeera and Mada Masr', *The Guardian*, 25 May, https://www.theguardian. com/world/2017/may/25/egypt-blocks-access-news-websites-al-jazeera-mada-masr-press-freedom (accessed 20 July 2017).

54. Rasha Abdulla, 2014, 'Egypt's media in the midst of revolution', *The Carnegie Endowment for International Peace*, 16 July, http://carnegieendowment. org/2014/07/16/egypt-s-media-in-midst-of-revolution-pub-56164 (accessed 7 June 2017).

55. Abolished in February of 2011, the Ministry of Information was reinstated just months later, on 12 July. See Committee to Protect Journalists, 2011, 'Egypt's reinstatement of Information Ministry is a setback', *CPJ*, 12 July, https://cpj.org/2011/07/egypts-reinstatement-of-information-ministry-is-ma. php (accessed 25 February 2018).

56. Abdallah F. Hassan, 2015, *Media, Revolution, and Politics in Egypt*, London: I. B. Tauris, p. 76.

57. See Emad Mekay, 2011, 'TV stations multiply as Egyptian censorship falls', *The New York Times*, 30 July, http://www.nytimes.com/2011/07/14/world/ middleeast/14iht-M14B-EGYPT-MEDIA.html (accessed 7 June 2017).

58. Hassan, 2015, p. 76.

59. Mekay, 2011.

60. Ibid.

61. Hamdi Qandil, 2014, *Ishtu marratayn*, Cairo: Dar al-Shorouk, p. 521.

62. Mamoun Fandi, 2008, *Hurub kalamiya*, London: Dar al-Saqi, p. 28.

63. See Armbrust, 2013.

64. Ismail al-Ashul, 2014, 'Hamdi Qandil', *Al-Shorouk*, February, http://www. shorouknews.com/news/view.aspx?cdate=04022014&id=67fe45ae-652c-42fd-afa8-dba8319af934 (accessed 25 February 2018).

65. Qandil, 2014, p. 521.

66. Al-Watan, 2014, 'Fadiha Fairmont: Safqa ma'a al-shaytan', *Al-Watan*, 17

February, http://www.elwatannews.com/news/details/419569 (accessed 21 July 2017).

67. Salma Shukrallah, 2013, 'Once election allies, Egypt's Fairmont opposition turn against Morsi', *Ahram Online*, 27 June, http://english.ahram.org.eg/ NewsContent/1/152/74485/Egypt/Morsi,-one-year-on/-Once-election-allies,-Egypts-Fairmont-opposition-.aspx (accessed 24 May 2017).

68. *Al-Watan*, 2014.

69. Ibid.

70. The exchange is available on the YouTube channel of *ONTV*, https://www. youtube.com/watch?v=CJCuDOEkLeo (accessed 9 October 2017).

71. Al-Jazeera published a transcript of the speech in Arabic, 2013, 'Nass bayan al-quwwat al-musallaha al-Misriya', *Al-Jazeera*, 2 July, http://www.aljazeera. net/news/reportsandinterviews/2013/7/2/%D9%86%D8%B5-%D8%A8% D9%8A%D8%A7%D9%86-%D8%A7%D9%84%D9%82%D9%88% D8%A7%D8%AA-%D8%A7%D9%84%D9%85%D8%B3%D9%84% D8%AD%D8%A9-%D8%A7%D9%84%D9%85%D8%B5%D8%B1% D9%8A%D8%A9-1-7-2013 (accessed 21 July 2017).

72. Ahmed al-Khatib, 2014, 'Kol ma yajib ta'rifuhu 'an 'hurub al-jil al-rabi'', *Sasapost*, 26 August, http://www.sasapost.com/4th-generation-warfare/ (accessed 21 July 2017).

73. Laurie Brand, 2014, *Official Stories: Politics and National Narratives in Egypt and Algeria*, Palo Alto: Stanford University Press, p. 102.

74. Lilia Blaise, 2014, 'De l'ATI à l'ATT: Quel avenir pour l'internet en Tunisie?', *Huffpost Maghreb*, 18 March, http://www.huffpostmaghreb.com/2014/03/18/ internet-tunisie_n_4985350.html (accessed 23 February 2018).

75. Magdi al-Galad, 2014, 'Infirad', *Al-Watan*, 1 June, https://www.elwatannews. com/news/details/495659 (accessed 25 February 2018).

76. Fatima Sharawai, 2015, 'Shabakat al-birnamij al-'amma', *Al-Ahram*, 16 September, http://gate.ahram.org.eg/News/751664.aspx (accessed 25 February 2018).

77. Sissela Bok, 1989, *Lying: Moral Choice in Public and Private Life*, New York: Vintage Books, p. 83.

78. Peter Pomerantsev, 2014, *Nothing is True and Everything is Possible*, New York: Public Affairs, p. 4.

79. On 'discoursive community', see Philippe-Joseph Salazar, 2017, *Words are Weapons: Inside ISIS's Rhetoric of Terror*, New Haven: Yale University Press, p. 155.

80. Alaa Al-Aswany, 2013, *The Automobile Club*, Cairo: Dar al-Shorouk, p. 239.

81. Amartya Kumar Sen, 2006, *Identity and Violence: the Illusion of Destiny*, New York: Norton, p. 46.

Chapter 5

1. RFI, 2015, 'Le groupe EI en Libye dit avoir tué deux journalistes tunisiens', *RFI*, 8 January, http://www.rfi.fr/afrique/20150108-branche-libyenne-ei-annonce-assassinat-chourabi-guetari/ (accessed 3 January 2016).

2. Kapitalis, 2015, 'Un témoin Libyen confirme l'assassinat de Sofiane et Nadhir', *Kapitalis*, 8 January, http://kapitalis.com/tunisie/2017/01/08/un-temoin-libyen-confirme-lassassinat-de-sofiane-et-nadhir/ (accessed 25 July 2017).

3. Ibid.

4. Mounira al-Bouti, 2016, 'La mère de Nadhir Ktari, disparu en Libye: "Notre visite a été fructueuse"', *Huffpost Maghreb*, 5 September, http://www.huff-postmaghreb.com/2016/09/05/sofiane-chourabi-nehdir-_n_11864758.html (accessed 30 January 2018).

5. Mosaique FM, 2017, 'Sofiane et Nédhir vivants: Bghouri émet des doutes', *Mosaique FM*, 16 July, http://www.mosaiquefm.net/fr/actualite-national-tunisie/172373/sofiane-et-nedhir-vivants-bghouri-emet-des-doutes (accessed 30 January 2018).

6. Foreign Policy, 2010, 'Think again: the Internet', *Foreign Policy*, 26 April, http://foreignpolicy.com/2010/04/26/think-again-the-internet/ (accessed 8 August 2017). An anonymous writer for the Muslim Brotherhood blog *Ikhwan Web* was also present at the event. He or she offers a list of the Middle Eastern participants. I have not been able to independently verify its accuracy. See http://www.ikhwanweb.com/article.php?id=23498 (accessed 8 August 2017).

7. See POMED, 2011, 'The Weekly Wire', *POMED*, 25 April, http://www.pomed.org/wp-content/uploads/2011/05/Weekly-Wire-April-25.pdf (accessed 8 August 2017).

8. A copy of Chourabi's talk is available online: https://www.youtube.com/watch?v=LHe0PdinBTE (accessed 8 August 2017).

9. Sami Ben Gharbia, 2010b, 'The Internet freedom fallacy and the Arab digital activism', *Nawaat*, 17 September, https://nawaat.org/portail/2010/09/17/the-internet-freedom-fallacy-and-the-arab-digital-activism/ (accessed 18 February 2018).

10. Yassine Ayari, 2011, 'Caid Essebsi, Marzouki, Rached Ghannouchi, Nejib Chebbi et la malaise de Yassine Ayari', *Mel7it* blog, 28 April, http://mel7it3.

blogspot.com/2011/04/caid-essebsi-marzouki-rached-ghannouchi.html
(accessed 23 July 2017).

11. Eric Goldstein and Oumayma Ben Abdallah, 2015, 'Missing journalists:
Tunisia's Arab Spring meets Libya's', *openDemocracy*, 5 May, https://www.
hrw.org/news/2015/05/05/missing-journalists-tunisias-arab-spring-meets-
libyas (accessed 30 January 2018).

12. Ibid.

13. Tunisia Live, 2012, 'The memories of journalist Sofiene Chourabi', *YouTube*,
18 January, https://www.youtube.com/watch?v=I8h5yv9BTf0 (accessed 15
February 2018).

14. Ibid.

15. Kristen Chick, 2011, 'How revolt sparked to life in Tunisia', *Christian
Science Monitor*, 23 January, https://www.csmonitor.com/World/Middle-
East/2011/0123/How-revolt-sparked-to-life-in-Tunisia (accessed 25 July
2017).

16. Mohamed Ghariani, the former Secretary General of the RCD became one
of the most prominent voices in promoting the false narrative of the '*cyber-
collabos*'. See Business News, 2018, 'Mohamed Ghariani: Des blogueurs et
des politiciens ont été formés pour renverser l'ancien régime!', *Business News*,
5 January, http://www.businessnews.com.tn/comment/874759 (accessed 15
February 2018).

17. Lina Ben Mhenni, 2017, '13 Mars', *Linkedin*, 17 March, https://fr.linkedin.
com/pulse/13-mars-lina-ben-mhenni (accessed 15 February 2018).

18. *Afrique Action* was renamed *Jeune Afrique* in 1961. The now renowned
journal, which covers all of Africa, remains one of the country's most vital
and independent media outlets. See Henry Clement Moore, 1973, 'Tunisia
after Bourguiba', in I. William Zartman (ed.), *Man, State, and Society in the
Contemporary Maghreb*, New York: Praeger, p. 273, and Moore, 1965, *Tunisia
Since Independence*, Berkeley: University of California Press, p. 221.

19. Salah al-Din al-Jourshi, 2011, 'Al-Mashahid al-Islami fi Tunis', in *Min Qabdat
Bin Ali ila thawrat al-yasmin*, Abu Dhabi: Al-Mesbar Studies & Research
Center, p. 26.

20. See, for example, Al-Jazeera, 2010, 'Tunis tahajaz sahifat al-mauqif al-mu'arida',
Al-Jazeera, 17 July, http://www.aljazeera.net/news/humanrights/2010/7/17/%
D8%AA%D9%88%D9%86%D8%B3-%D8%AA%D8%AD%D8%AC%
D8%B2-%D8%B5%D8%AD%D9%8A%D9%81%D8%A9-%D8%A7%
D9%84%D9%85%D9%88%D9%82%D9%81-%D8%A7%D9%84%

D9%85%D8%B9%D8%A7%D8%B1%D8%B6%D8%A9 (accessed 16 February 2018).

21. See Mohamed Zayani, 2015, *Networked Publics and Digital Contention: The Politics of Everyday Life in Tunisia*, Oxford: Oxford University Press, pp. 95–7. *Tunisnews*, which Zayani covers extensively, is an online publication not bound to Ennahdha or any other political party. As he notes, the website publishes material from a range of voices and is not exclusively Islamist in its orientation.

22. For an excellent and detailed discussion of online publishing in the early 2000s, see Amy Aisen Kallander, 2013, 'From TUNeZINE to Nhar 3la 3mmar: a reconsideration of the role of bloggers in Tunisia's revolution', *Arab Media & Society*, 27 January http://www.arabmediasociety.com/?article=818 (accessed 17 February 2018). Also, Zayani, 2015, p. 98. As Zayani notes, public access to the Internet was not provided until 1996 (p. 80).

23. Zayani, 2015, p. 100.

24. See Frida Dahman, 2015, 'Tunisie: le Mouvement du 18 octobre 2005, 10 ans après', *Jeune Afrique*, 28 October, http://www.jeuneafrique.com/274962/politique/tunisie-le-mouvement-du-18-octobre-2005-10-ans-apres/.

25. Zayani, 2015, p. 104.

26. The Yezzi Fock Mission Statement is available on Facebook: https://web.facebook.com/pg/yezzifock/about/?ref=page_internal (accessed 4 August 2017).

27. See the *Nawaat* website at http://nawaat.org/portail/.

28. Zayani, 2015, p. 103.

29. The French-based *Nawaat* member 'Astrubal' used the data to argue the President and his entourage had been using the airplane for illicit 'shopping sprees in Europe'. See Zayani, 2015, pp. 104–5.

30. Fred Forest, 1983, *Manifeste pour une esthetique de la communication*, Brussels: Revue D'Art Contemporain, http://www.webnetmuseum.org/html/fr/expo-retr-fredforest/textes_critiques/textes_divers/3manifeste_esth_com_fr.htm#text (accessed 13 February 2018).

31. Martin Heidegger, 1977, *Basic Writings*, New York: Harper and Row, p. 157.

32. See Sami Ben Gharbia, 2010a, 'Anti-censorship movement in Tunisia: creativity, courage and hope!', *Global Voices*, 27 May, https://advox.globalvoices.org/2010/05/27/anti-censorship-movement-in-tunisia-creativity-courage-and-hope/print/.

33. See Minbar al-Jazeera, 2010, 'Hajb al-mawaqi' al-iliktruniya', *Al-Jazeera*, 24 May, http://www.aljazeera.net/programs/al-jazeera-platform/2010/5/24/%D8%AD%D8%AC%D8%A8-%D8%A7%D9%84%D9%85%D9%88

D8%A7%D9%82%D8%B9-%D8%A7%D9%84%D8%A5%D9%84%
D9%83%D8%AA%D8%B1%D9%88%D9%86%D9%8A%D8%A9
(accessed 17 February 2018).

34. See Azyz Amami, 2012, 'My Palestine', *Al-Arabiya English*, 22 November, https://english.alarabiya.net/authors/Azyz-Amami.html (accessed 20 February 2013).

35. See Nawaat, 2010, 'TuniLeaks, les documents dévoilés par WikiLeaks concernant la Tunisie: Quelques réactions à chaud', *Nawaat*, 28 November, http://nawaat.org/portail/2010/11/28/tunileaks-les-documents-devoiles-par-wikile aks-concernant-la-tunisie-quelques-reactions-a-chaud/ (accessed 17 February 2018).

36. Sami Ben Gharbia, 2017, 'Shahadat Sami Ben Gharbia', *Instance Vérité & Dignite Commission, Tunis*, 11 March, https://www.youtube.com/watch?v=ZOC3Rp9Zvks (accessed 1 September 2017).

37. See Tarek Kahlaoui, 2013, 'The powers of social media', in Nouri Gana (ed.), *The Making of the Tunisian Revolution*, Edinburgh: Edinburgh University Press, p. 151. Also, Zayani, 2015, p. 161.

38. Ben Gharbia, 2010a. In the lead up to anti-censorship rallies scheduled for 22 May 2010, she had her home broken into and her computer stolen. As she told Kristen McTighe for *The New York Times* 'it was clear it was them [i.e. the police] because of the way only I was targeted and the way they went after my equipment'. See Kristen McTighe, 2011, 'A blogger at Arab Spring's genesis', *The New York Times*, 12 October, http://www.nytimes.com/2011/10/13/world/africa/a-blogger-at-arab-springs-genesis.html.

39. Zayani, 2015, p. 162.

40. Sami Ben Gharbia, 2012, 'Curating reality: new tools for investigative journalism', Sandberg @mediafonds conference, Amsterdam, Netherlands, 9 February, https://vimeo.com/36764899 (accessed 11 February 2018).

41. Le Monde, 2011, 'L'Egypt: autre "ennemi d'Internet"', *Le Monde*, 26 January, http://www.lemonde.fr/technologies/article/2011/01/26/l-egypte-autre-ennemi-d-internet_1470630_651865.html (accessed 13 July 2017).

42. Reporters without Borders, 2002, 'Imprisoned cyber-dissident risks five-year jail sentence', *Reporters without Borders*, 12 June, https://rsf.org/en/news/imprisoned-cyber-dissident-risks-five-year-jail-sentence (accessed 24 July 2017).

43. Dermott McGrath, 2002, 'Tunisian net dissident jailed', *Wired*, 13 June, https://www.wired.com/2002/06/tunisian-net-dissident-jailed/ (accessed 24 July 2017).

44. Kahlaoui, 2013, p. 150.
45. Rebecca MacKinnon, 2013, *Consent of the Networked: The Worldwide Struggle For Internet Freedom*, New York: Basic Books, p. 4.
46. Marketwired, 2010, 'Narus wins award for best real-time dynamic network analysis and forensics solution', *Marketwired*, 9 November, http://www.marketwired.com/press-release/narus-wins-award-for-best-real-time-dynamic-network-analysis-and-forensics-solution-nyse-ba-1349435.htm (accessed 7 February 2018).
47. Ben Gharbia, 2017.
48. Ben Gharbia, 2012.
49. Ibid.
50. See Ellery Roberts Biddle and Afef Abrougui, 2017, 'Tunisian media activist interrogated over sources of leaked documents', *Global Voices*, 8 May, https://advox.globalvoices.org/2017/05/08/tunisian-media-activist-interrogated-over-sources-of-leaked-documents/ (accessed 14 February 2018).
51. Ibid.
52. Sami Ben Gharbia, 2003, *Borj Erroumi XL: voyage dans un monde hostile*, https://ifikra.files.wordpress.com/2009/12/borjerroumixl.pdf (accessed 24 February 2018).
53. See André Gunthert, 2011, 'Photographier la révolution tunisienne', Mp3 recording with Azyz Amami, *L'Ateliêr des icons*, http://histoirevisuelle.fr/cv/icones/1860.
54. Marwan Kraidy, 2017, *The Naked Blogger of Cairo*, Cambridge, MA: Havard University Press, p 17.
55. Ben Gharbia, 2012.
56. See Mounir Troudi and Ammar404tounsi, 2010, YouTube, 25 May, https://www.youtube.com/watch?v=YRZipuMCxkI (accessed 6 February 2018).
57. See Néstor García Canclini, 1995, *Hybrid Cultures: Strategies for Entering and Leaving Modernity*, Minneapolis: University of Minnesota Press, p. 91.
58. Koldo Gutiérrez, 'Arte desde la trinchera en apoyo a los maestros de Oaxaca', *Cactus*, https://www.revistacactus.com/arte-desde-la-trinchera-en-apoyo-a-los-maestros-de-oaxaca/ (accessed 26 February 2018).
59. Ben Gharbia, 2012.
60. Canclini, 1995, p. 90. Author's emphasis.
61. Ibid. p. 89.
62. The photo's caption reads: 'A call (*da'wa*) to the neighborhood committees to reunite in all districts and to confront saboteurs with full force ... The

saboteurs and supporters of the former regime compel their ranks to destabilize the country, to create bloodshed and create confusion as a means of polluting the gains of the revolution . . . So we must stand as one hand (*eid wahda*) to protect our people and property . . . Long live the people and may the cowards' eyes find no rest.'

63. The accompanying post reads: 'No to the Republic of Blocking, yes to Internet Freedom.'

64. Al-Jubha al-tahyes al-sha'biya, 2011, 'Libya 2.0', 17 February, Tahyess.blogspot.com (accessed 26 June 2018).

65. Jim Sisco, 2015, 'Should the Gulf states be weary of the revolution tsunami coming their way', *The Express Tribune*, 23 August, https://blogs.tribune.com.pk/story/29134/should-the-gulf-states-be-wary-of-the-revolution-tsunami-coming-their-way/ (accessed 20 February 2018).

66. Quoted in Zayani, 2015, p. 104.

67. Dean Wright, 2009, 'The fall of the wall – and the media's role', *Reuters*, 9 November, http://blogs.reuters.com/fulldisclosure/2009/11/09/the-fall-of-the-wall-and-the-medias-role/ (accessed 30 July 2017).

68. Zayani, 2015, p. 99.

69. Ben Gharbia, 2003.

70. For the history of *Global Voices*, see MacKinnon, 2012, pp. xii–xiii. As MacKinnon describes, *Global Voices* developed from a conference blog in 2004. Among the co-collaborators she identifies at the early stage was Ethan Zuckerman, a well-known media scholar and prolific blogger.

71. Michael Young, 2007, *Death, Sex & Money: Life Inside a Newspaper*, Carlton: Melbourne University Press, p. 191.

72. Wright, 2009.

73. Jean-Pierre Filiu, 2011, *The Arab Revolution: Ten Lessons from the Democratic Uprising*, New York: Oxford University Press, p. 176 n. 20. Filiu provides no documentation for this claim though I have seen no evidence to dispute it.

74. Lina Ben Mhenni, 2010, 'Tunisia: Unemployed man's suicide attempt sparks riots', *Global Voices*, 23 December, https://globalvoices.org/2010/12/23/tunisia-unemployed-mans-suicide-attempt-sparks-riots/ (accessed 1 February 2018).

75. János László, 2008, *The Science of Stories: An Introduction to Narrative Psychology*, New York: Routledge, p. 142.

76. Ibid. p. 141.

77. Ibid. p. 141.

78. Quoted in Sam Cherribi, 2017, *Fridays of Rage: Al-Jazeera, the Arab Spring, and Political Islam*, Oxford: Oxford University Press, p. 84.

79. Lina Ben Mhenni, 2011, *Tunisian Girl/Bunaiyat Tunissiya: Blogueuse pour un printemps arabe*, Montpellier: Indigène, p. 12.

80. Zuheir Makhlouf, 2011, 'Muhawala intihar al-sha'b al-majaz Mohamed al-Bou'azizi asil Sidi Bouzid', *Assabilonline*, 17 December, https://www.facebook.com/168736489831044/videos/119774181422660/ (accessed 29 June 2018).

81. France 24, 'Muwajahat bein al-shurta wa muhtajjin fi wilayat Sidi Buzid', *France 24*, 19 December, http://www.france24.com/ar/20101219-tunisia-clash-poorness-young-suicide-police (accessed 17 March 2017).

82. See Zayani, 2015, p. 70 n. 45. In a different context, Zayani supplies the title of Horchani's article. The note is unrelated to Horchani's role as a witness for the *Reuters* report.

83. See Sara Ledwith, 2009, 'Iran's Neda shows citizen journalism unleashed', *Reuters*, 23 June, https://www.reuters.com/article/us-internet-iran-analysis-sb/irans-neda-shows-citizen-journalism-unleashed-idUKTRE55M39820090623 (accessed 18 June 2018). See also 'Video shows Iranian woman bleeding to death in Tehran', *The Telegraph*, 21 June 2009, http://www.telegraph.co.uk/news/worldnews/middleeast/iran/5595701/Video-shows-Iranian-woman-bleeding-to-death-in-Tehran.html (accessed 6 August 2017). The original video, published 20 June 2009, can be found at: https://www.youtube.com/watch?v=ehtIZQypu_w. In the Iranian media sphere, needless to say, the incident has been the subject of relentless conspiracy theory. Dean Wright has denied any connection between the State Department's Public Diplomacy 2.0 campaign and *Reuters* (personal interview with the author, 2018).

84. Slimane Rouissi, 2012, 'Connaissez-vous Slimane Rouissi: l'homme qui a lance la revolution Tunisienne', *France 24*, 17 December, http://observers.france24.com/fr/20111216-connaissez-vous-slimane-rouissi-l%E2%80%99homme-la-nce-revolution-tunisienne-sidi-bouzid-bouazizi-14-janvier (accessed 6 August 2017).

85. Slimane Rouissi, 2010, 'Public suicide sparks angry riots in central Tunisia', *The Observers*, 21 December, http://observers.france24.com/en/20101221-youth-public-suicide-attempt-sparks-angry-riots-sidi-bouzid-tunisia-poverty-bouazizi-immolation (accessed 6 August 2017).

86. Cherribi, 2017, p. 81.

87. Ibid. p. 81.

88. Merlyna Lim, 2013, 'Framing Bouazizi: "White lies", hybrid network, and collective/connective action in the 2010–11 Tunisian uprising', *Journalism* 14.7.

89. Ben Mhenni, 2011, p. 12.

90. Ahmed Mansour, 2012, 'Lina Ben Mhenni. Shahada 'ala al-thawra al-Tunissiya', *Al-Jazeera*, 25 July, http://www.aljazeera.net/programs/century-witness/2012/7/25/%D9%84%D9%8A%D9%86%D8%A7-%D8%A8%D9%86-%D9%85%D9%87%D9%86%D8%A7-%D8%B4%D8%A7%D9%87%D8%AF%D8%A9-%D8%B9%D9%84%D9%89-%D8%A7%D9%84%D8%AB%D9%88%D8%B1%D8%A9-%D8%A7%D9%84%D8%AA%D9%88%D9%86%D8%B3%D9%8A%D8%A9-%D8%AC2 (accessed 19 February 2018).

91. Ben Mhenni, 2011, p. 4.

92. Global Voices, 'Lina Ben Mhenni', *Global Voices*, https://globalvoices.org/author/lina-ben-mhenni/.

93. See Roman Lecomte, 2009, 'Internet et la reconfiguration de l'espace public tunisien: le rôle de la diaspora', *Nawaat*, 20 December, https://nawaat.org/portail/2009/12/20/internet-et-la-reconfiguration-de-lespace-public-tunisien-le-role-de-la-diaspora/ (accessed 24 July 2017).

94. Ben Gharbia, 2012.

95. Lina Ben Mhenni, 2008, 'Tunisia: National day for freedom of blogging on November 4', *Global Voices*, 10 October, https://globalvoices.org/2008/10/10/tunisia-national-day-for-freedom-of-blogging-on-november-4/ (accessed 26 February 2018). The site appeared blocked for approximately two weeks according to Ben Mhenni and others. The reason for its shut down was presumably to limit the dissemination of news surrounding the ongoing strikes in Gafsa.

96. Sami Ben Gharbia, 2008, 'Tunisia seems to have blocked access to Facebook', *Global Voices*, 18 August, https://advox.globalvoices.org/2008/08/18/tunisia-seems-to-have-blocked-access-to-facebook/.

97. Ben Mhenni, 2008.

98. The Observers, 2013, 'Tunisian hackers decrypt dictator's old Internet censorship machines', *France 24*, 24 June, http://observers.france24.com/en/20130624-tunisia-internet-censorship-hackers-servers (accessed 24 July 2017).

99. Yasmine Ryan, 2011, 'Tunisia's bitter cyberwar', *Al-Jazeera*, 6 January, http://www.aljazeera.com/indepth/features/2011/01/20111614145839362.html (accessed 6 January 2012).

100. Isaiah Berlin, 1958, 'The Origins of Israel', in Walter Z. Laqueur (ed.), *The Middle East in Transition*, New York: Praeger, p. 207.

Chapter 6

1. Jacques Rancière, 2007, *On the Shores of Politics*, London: Verso, p. 11.
2. On the occasion of the fifty-seventh anniversary of independence, Rached al-Ghannouchi, the leader of Ennahdha, issued a statement declaring: 'The 20th of March 1956 represented a critical date in the history of building the modern national state, and will remain a symbolic day in the history and memory of Tunisians, despite the deviations of this experience and the transformation of the country under the deposed former president to a center of authoritarianism and corruption against which the Tunisian people revolted, toppling tyranny and its symbols.' See Harakat al-Nahdha, 2013, 'Tadaʿwa ila Rachid', *Harakat al-Nahdha*, 20 March, http://www.ennahdha.tn/ (accessed 20 March 2013).
3. Ettakatol website, 2013, http://ettakatol.org/?lang=fr (accessed 1 May 2013).
4. Nouri Gana, 2017, 'Sons of a beach: the politics of bastardy in the cinema of Nouri Bouzid', *Cultural Politics*, 13.2, 177–93.
5. See Yassine Ayari, 2011, 'Caïd Essebsi, Marzouki, Rached Ghannouchi, Nejib Chebbi et la malaise de Yassine Ayari', *Mel7it*, 28 April, http://mel7it3.blogspot.com/2011/04/caid-essebsi-marzouki-rached-ghannouchi.html (accessed 23 July 2017). Ayari was arrested himself in 2014 for 'defaming the military', his indictment the result of the then recently created 'Law of Repression of Attacks Against the Armed Services'. As the incident showed, deference to the armed services, unlike under the rule of Ben Ali, would be enforced across the social spectrum. Not only was Ayari not an Islamist militant, but his father, as reported by Human Rights Watch, had been assassinated in 2011, presumably by a Salafi jihadist.
6. Youssef Seddik, 2011, 'La Palestine, notre citoyenne . . .', *Le Temps*, 10 December, http://www.letemps.com.tn/article-61455.html (accessed 13 September 2017).
7. Youssef Seddik, 2014, *Tunisie: La revolution inachevée. Entretiens avec Gilles Vanderpooten*, Paris: Editions de l'Aube, p. 72.
8. Youssef Seddik, 2011, 'Indépendance', *Le Temps*, 6 December, http://www.letemps.com.tn/article-61304.html (accessed 15 September 2017).
9. Youssef Seddik, 2013, 'Figures d'une Révolution qui n'a pu devenir . . .', *Le Temps*, 24 January, http://www.letemps.com.tn/article-61304.html (accessed 15 September 2017).

10. Ibid.
11. Youssef Seddik, 2013, *Nous n'avons jamais lu le Coran*, Paris: l'Aube, pp. 36–7.
12. Seddik, 2014, p. 72.
13. Ibid. p. 73.
14. Youssef Seddik, 2012, 'Le sang de Monsieur Abdelfattah Mourou', *Le Temps*, 11 August, http://www.letemps.com.tn/article-61304.html (accessed 1 September 2012).
15. Youssef Seddik, 2011, 'Indépendance', *Le Temps*, 6 December, http://www. letemps.com.tn/article-61304.html (accessed 13 September 2017).
16. Raymond Williams uses the phrase 'structure of feelings' throughout his work. I take its meaning here from his usage in *Marxism and Literature* where he explains it in relation to the role of literature as a formal medium of 'individual variations' designed for the articulation of a social moment. See R. Williams, 1977, *Marxism and Literature*, Oxford: Oxford University Press, p. 191.
17. Youssef Seddik, 2012, 'Encore une lettre à un(e) inconnu(e) intelligent(e) parmi nos gouvernants provisoires', *Le Temps*, 25 August, http://www.letemps. com.tn/article-69172.html (accessed 1 July 2013).
18. See Christophe Cotteret, 2014, *Ennahda, un histoire Tunisienne* [documentary], ARTE, http://www.tv5monde.com/programmes/fr/programme-tv-ennahdha-une-histoire-tunisienne/22801/ (accessed 28 February 2018).
19. Seddik, 2014, p. 29.
20. Azzam Tamimi, 2001, *Rached Ghannouchi: A Democrat within Islamism*, New York: Oxford University Press, p. 6.
21. Youssef Seddik, 2016a, *Le grand malentendu: l'Occident face au Coran*, La Tour-d'Aigues: Editions de l'Aube, p. 31.
22. Ibid. p. 32.
23. Youssef Seddik, 2016b, 'Al-Mufakkir al-Tunisi Youssef Seddik', *Al-Araby TV*, 7 May, https://www.youtube.com/watch?v=geHa5fRUSWA (accessed 8 September 2017).
24. According to Ghannouchi's own account, 15 June 1966 was the night in Syria that he converted to Islamism. 'That very night I shed two things off of me: secular nationalism and traditional Islam', he later recalled in an interview with Azzam S. Tamimi. See Tamimi, 2001, p. 22.
25. Rached al-Ghannouchi, 2005, *Al-Qadr 'anda Ibn Taymiyya*, Saudi Arabia: al-Markaz al-Raya, p. 19.
26. Ibid. p. 23.
27. Quoted in Cotteret, 2014.

28. Rached al-Ghannouchi, 1979, 'Al-Thawra al-Iraniyya thawra Islamiyya', *Al-Ma'rifa*, 12 February, 3.2.

29. Rached al-Ghannouchi and Olivier Ravanello, 2015, *Entretiens d'Olivier Ravanello avec Rached Ghannouchi*, Paris: Plon, p. 99.

30. Paul Ricoeur, 2005, *The Course of Recognition*, Cambridge, MA: Harvard University Press, pp. 101–2.

31. Youssef Seddik, 2011, 'La chose publique – "Notre Constitution" . . . naissance d'un réseau', *Le Temps*, 29 July, http://www.letemps.com.tn/article-57918. html (accessed 20 September 2017).

32. Ibid.

33. The Tunisian Constitution of 2014 can be found online at https://www.constituteproject.org/constitution/Tunisia_2014.pdf.

34. Jeffry R. Halverson and Nathaniel Greenberg, 2018, *Islamists of the Maghreb*, London: Routledge, p. 83.

35. Rached al-Ghannouchi, 1993, *Al-Hurriyat al-'amma fi al-dawla al-Islamiya*, Beirut: Merkaz Darasat al-Wahda al-'Arabiyya, p. 319.

36. Al-Ghannouchi and Ravanello, 2015, p. 99.

37. Reference to *ijtihad* infused virtually all aspects of the Islamists' political negotiations in the post-revolutionary context. Kennith Perkins notes, for example, during the 2011 Constituent Assembly elections, Ghannouchi cited *ijtihad* as 'cause enough' for avoicing the annulment of the Personal Status Code which granted women rights comparable to men. See Kenneth Perkins, 2013, 'Playing the Islamist card: the use and abuse of religion in Tunisian politics', in Nouri Gana (ed.), *The Making of the Tunisian Revolution*, Edinburgh: Edinburgh University Press, p. 61.

38. Al-Ghannouchi and Ravanello, 2015, p. 100.

39. See Jean Wahl, 1926, *Étude sur le Parménide de Platon*, Paris: F. Rieder & Cie.

40. Jean Wahl, 2005, *Les philosophies pluralistes d'Angleterre et d'Amérique*, Paris: Les empêcheurs de penser en rond, p. 212.

41. Youssef Seddik, 2005, *Qui sont les barbares? Itinéraire d'un penseur d'islam*, Paris: Editions l'Aube, p. 35.

42. Jorge Luis Borges, 1977, *El libro de arena*, Buenos Aires: Biblioteca Borges, p. 19.

43. Youssef Seddik, 2011, 'Foi et bonne foi', *Le Temps*, 17 December, http://www.letemps.com.tn/article-61657.html (accessed 12 September 2017).

44. Seddik, 2014, p. 148.

45. Ibid. p. 52.

46. Seddik, 2005, p. 35.
47. Seddik, 2013, p. 78.
48. Sophie Rostain, 2005, 'L'Islam à Coran ouvert', *Libération*, 22 April, http://www.liberation.fr/medias/2005/04/22/l-islam-a-coran-ouvert_517419 (accessed 8 September 2017).
49. See Tawasol, 2012, 'Turida Youssef Seddik min jamaʿat al-Zaytuna', *Tawasol TV*, 26 April, https://www.youtube.com/watch?v=-Zgp5MwM9QM (accessed 8 September 2017).
50. See Huffpost Maghreb, 2015, 'Tunisie: Le président du Conseil islamique révoqué après avoir comparé Youssef Seddik à Salman Rushdie', *Huffpost Maghreb*, 7 April, http://www.huffpostmaghreb.com/2015/07/04/tunisie-conseil-islamique-youssef-seddik_n_7728078.html (accessed 8 September 2017).
51. Youssef Seddik, 2012, 'Ghannouchi, les laïcs et les "athées"', *Le Temps*, 7 March, http://www.letemps.com.tn/article-64082.html (accessed 14 September 2017).
52. Ibid.
53. Al-Ghannouchi, 2005, p. 23.
54. Ibid. p. 19.
55. Nadia Marzouki, 2013, 'From resistance to governance: the category of civility in the political theory of Tunisian Islamists', in Nouri Gana (ed.), *The Making of the Tunisian Revolution: Contexts, Architects, Prospects*, Edinburgh: Edinburgh University Press, p. 211.
56. Just a month after his interview on *Al-Jazeera*, a leaked video surfaced on YouTube featuring Ghannouchi telling two Salafi activists to 'be patient'. Sounding much like his more radical counterparts, he emphasised the recent history of Algeria. 'We don't want to go down their path', he said, referring to the overthrow of the Islamist government in 1991 and the brutal civil war that ensued. 'Today we have more than a mosque', he said, 'we have a Ministry of Religious Affairs. We have more than a single store. We have a State.' The video can be found at http://www.youtube.com/watch?v=Qu2TXVzQXQ4. See also, Marc Daou, 2012, 'Filmé à son insu, Rached Ghannouchi tombe la masque', *France 24*, 11 October, http://www.france24.com/fr/20121011-tunisie-rached-ghannouchi-islamisme-salafistes-video-laics-polemique-ennahda (accessed 11 January 2018).
57. Marzouki, 2013, p. 211.
58. Shaykh Abu-Muslim al-Jazairi, 2011, 'Fatwa supports "Campaign to defend the

Islamic identity of Tunisia"', *Al-Andalus Establishment for Media Production*, http://www.opensource.gov (accessed 18 May 2013).

59. Seddik, 2014, p. 193.
60. Ibid. p. 193.
61. Ibid. p. 185.
62. János László, 2008, *The Science of Stories: An Introduction to Narrative Psychology*. New York: Routledge, p. 165.
63. Youssef Seddik, 2011, 'La Palestine, notre citoyenne', *Le Temps*, 10 December, http://www.letemps.com.tn/article-61455.html (accessed 1 October 2013).
64. The speech, titled '*Taharrur al-shu'ub al-'Arabiya min al-istibdad khatwa 'ala tariq tahrir Filistin*' (The liberation of the Arab Peoples: stepping stones to the liberation of Palestine) was posted on the group's Facebook page and party website on 17 May 2012, but later removed. A copy of the remarks can be found at *Turess.com*, http://www.turess.com/infosplus/11450 (accessed 1 September 2016).
65. Ghannouchi and Ravanello, 2015, p. 107.
66. Tamimi, 2001, p. 113.
67. From Karl Marx, *The Eighteenth Brumaire of Louis Napoleon* (1852).

Chapter 7

1. Human Rights Watch, 2011, 'Tunisia: Hold police accountable for shootings', *Human Rights Watch*, 29 January, https://www.hrw.org/news/2011/01/29/tunisia-hold-police-accountable-shootings (accessed 26 February 2018).
2. Al-Jazeera, 2011, 'Khitab Ben Ali', *Al-Jazeera*, 14 January, http://www.aljazeera.net/news/arabic/2011/1/14/%D8%AE%D8%B7%D8%A7%D8%A8-%D8%A8%D9%86-%D8%B9%D9%84%D9%8A-%D9%8A%D8%AB%D9%8A%D8%B1-%D8%B1%D8%AF%D9%88%D8%AF-%D8%A3%D9%81%D8%B9%D8%A7%D9%84-%D9%85%D8%AA%D9%86%D8%A7%D9%82%D8%B6%D8%A9 (accessed 14 June 2017).
3. Ibid.
4. Al-Andalus Establishment for Media Production, 2011a, 'Al-Andalus releases audio by AQLIM leader on recent unrest in Tunisia' ['Help and support for our people's uprising in Tunisia'], *Al-Andalus Establishment for Media Production*, 13 January, http://www.opensource.gov (accessed October 2012).
5. Al-Andalus Establishment for Media Production, 2014, *al-Jaza'ir wa al-nafaq al-muzlim*, https://www.youtube.com/watch?v=27iTIWAin78&t=21s (accessed 15 June 2017).

6. Paul Ricoeur, 2005, *The Course of Recognition*, Cambridge, MA: Harvard University Press, pp. 101–2.

7. AQIM Media Committee, 2009, 'AQLIM: "Declaration of al-Andalus Establishment for Media Production"' [Good news: Declaration of al-Andalus Establishment for Media Production], *Al-Andalus Establishment for Media Production*, 4 October, http://www.opensource.gov (accessed October 2012).

8. Kamal Zayit, 2014, 'Al-jamaʿat al-musallaha fi Jazaʾir min "Buyaʿli' ila 'jund al-khilafa"', *Al-Quds al-Arabi*, 11 October, http://www.alquds.co.uk/?p=233522 (accessed 2 June 2017).

9. Nouri al-Jarrah, 2000, *Al-Firdaws al-dami*, Beirut: Riad el-Rayyes, p. 83.

10. See Alison Pargeter, 2012, 'Radicalisation in Tunisia', in George Joffe (ed.), *Islamist Radicalisation in North Africa: Politics and Process*, London: Routledge.

11. Zayit, 2014.

12. Nouri Al-Jarrah, 1998, 'Al-ʿAzima fi sanatiha al-ʿashara', *Al-Hayat*, 5 May, http://www.alhayat.com/article/963279/%D8%A7%D9%84%D8%A3%
D8%B2%D9%85%D8%A9-%D9%81%D9%8A-%D8%B3%D9%86%
D8%AA%D9%87%D8%A7-%D8%A7%D9%84%D8%B9%D8%A7%
D8%B4%D8%B1%D8%A9-%D8%A7%D9%84%D8%AD%D9%
8A%D8%A7%D8%A9-%D8%AA%D8%AA%D8%AC%D9%88%
D9%84-%D8%B9%D9%84%D9%89-%D9%85%D8%AF%D8%A7%
D8%B1-31-%D9%8A%D9%88%D9%85%D8%A7-%D9%81%D9%
8A-%D8%A7%D9%84%D8%AC%D8%B2%D8%A7%D8%A6%D8%
B1-10-%D8%B1%D8%B3%D8%A7%D9%85-%D8%A7%D9%84%
D9%83%D8%A7%D8%B1%D9%8A%D9%83%D8%A7%D8%AA%
D9%88%D8%B1-%D8%A3%D9%8A%D9%88%D8%A8-%D8%A7%
D9%84%D9%85%D8%BA%D8%B6%D9%88%D8%A8-%D8%B9%
D9%84%D9%8A%D9%87-%D9%85%D9%86-%D8%A7%D9%84%
D8%B3%D9%84%D8%B7%D8%A9-%D8%A7%D9%84%D8%B5%
D8%AD%D8%A7%D9%81%D8%A9-%D8%A7%D9%84%D8%AC%
D8%B2%D8%A7%D8%A6%D8%B1%D9%8A%D8%A9-%D9%85%
D8%AA%D9%88%D8%A7%D8%B7%D8%A6%D8%A9-%D9%88%
D9%87%D9%8A-%D8%AA%D8%B9%D8%B1%D9%81-%D9%85%
D9%86-%D8%A7%D9%84%D8%B0%D9%8A-%D9%8A%D9%82%
D8%AA%D9%84-!-nbsp (accessed 3 May 2017).

13. Jean-Pierre Filiu, 2009, 'Al-Qaeda in the Islamic Maghreb: Algerian challenge or global threat?', *Carnegie Papers* 104, quoted in Jeffry R. Halverson

and Nathaniel Greenberg, 2017, 'Ideology as narrative: the mythic discourse of al-Qaeda in the Islamic Maghrib', *Middle East Journal of Culture and Communication* 10.1.

14. AQIM Media Committee, 2009.
15. Abu Obayda Yusuf al-Annabi, 2011, ['An interview with Abu Ubaydah Yusuf al-Annabi'], *Al-Andalus Establishment for Media Production*, 7 July, http://www.opensource.gov (accessed October 2012).
16. Al-Andalus Foundation for Media Production, 2014.
17. Al-Andalus Establishment for Media Production, 2011a.
18. For the latter, this element of critique was engrained in its platform. The UGTT adopted a more aggressive stance towards the RCD following the first Gulf War, according to Sami Zemni. See Zemni, 2013, 'From socio-economic protest to national revolt: the labor origins of the Tunisian Revolution', in Nouri Gana (ed.), *The Making of the Tunisian Revolution*, Edinburgh: Edinburgh University Press, p. 139.
19. See Al-Jazeera, 2014, 'Harakat al-tawhid wa jihad fi gharb Afriqiya', *Al-Jazeera*, http://www.aljazeera.net/encyclopedia/movementsandparties/2014/2/12/%D8%AD%D8%B1%D9%83%D8%A9-%D8%A7%D9%84%D8%AA%D9%88%D8%AD%D9%8A%D8%AF-%D9%88%D8%A7%D9%84%D8%AC%D9%87%D8%A7%D8%AF-%D9%81%D9%8A-%D8%BA%D8%B1%D8%A8-%D8%A3%D9%81%D8%B1%D9%8A%D9%82%D9%8A%D8%A7 (accessed 9 January 2018).
20. Abdullah Moulouc, 2017, 'Aymin 'Aam "harakat tahrir Azawad"', *Al-Quds al-'arabi*, 13 May, http://www.alquds.co.uk/?p=718667 (accessed 9 January 2018).
21. See Aqlame, 2014, 'I'lan madinat Ghao imara tabi'a l-il-da'ish', *Aqlame.com*, http://www.aqlame.com/article19481.html (accessed 9 January 2018).
22. Quoted in Mathieu Guidère, 2014, 'The Timbuktu letters: new insights about AQIM', *Res Militaris* 4.1, 7.
23. Fredric Jameson, 1991, *Postmodernism, or The Culture Logic of Late Capitalism*, Durham, NC: Duke University Press, p. 27.
24. Guidère, 2014, p. 3.
25. See Jameson, 1991, pp. 26–7.
26. Ibid. p. 27.
27. See Halverson and Greenberg, 2017.
28. Al-Annabi, 2011.
29. Ibid.

30. Ibid.

31. Ibid.

32. Hani Shukrallah, 2012, 'Covering the Arab Spring: myths, lies, and truths', Issam Ferres Institute for Public Policy and International Affairs, *The American University in Beruit*, 21 May, https://www.youtube.com/watch?v=Wu3nwHhAI2U (accessed 1 June 2016).

33. Al-Annabi, 2011.

34. For a translation of al-Shabbi's poem see Gael Raphael, 2011, 'Al-Shaabi's "The Will to Life"', *Jadaliyya*, 2 May, http://www.jadaliyya.com/Details/23935/Al-Shabbi%60s-The-Will-to-Life (accessed 11 January 2018).

35. The video can be found at http://www.youtube.com/watch?v=Qu2TXVzQXQ4. See also Marc Daou, 2012, 'Filmé à son insu, Rached Ghannouchi tombe la masque', *France 24*, 11 October, http://www.france24.com/fr/20121011-tuni sie-rached-ghannouchi-islamisme-salafistes-video-laics-polemique-ennahda (accessed 11 January 2018).

36. Al-Annabi, 2011.

37. Al-Andalus Establishment for Media Production, 2012a, ['In the fields and squares of liberation . . .'], *Al-Andalus Establishment for Media Production*, 22 May, http://opensource.gov (accessed October 2012).

38. See Hans Robert Jauss, 1982, *Toward an Aesthetic of Reception*, trans. Timothy Bahti, Minneapolis: University of Minnesota Press, p. x.

39. Ricoeur, 2005, p. 102.

40. Ibid. p. 100.

41. Al-Andalus Establishment for Media Production, 2012a.

42. Ibid.

43. See Marc Sageman, 2004, *Understanding Terror Networks*, Philadelphia: University of Philidelphia Press, *passim*.

44. Ibid. p. 36.

45. Jameson, 1991, p. 18.

46. Sageman provides a sample breakdown of the socioeconomic status of core al-Qaeda members in his 2004 study, pp. 73–4.

47. François Burgat and William Dowell, 1997, *The Islamic Movement in North Africa*, Austin: Center for Middle Eastern Studies, University of Texas at Austin, p. 22.

48. Jameson, 1991, p. 20.

49. Ibid. p. 27.

50. Al-Andalus Establishment for Media Production, 2012b, ['Our Muslim broth-

ers in Tunisia'], 22 October, *Al-Andalus Establishment for Media Production*, http://opensource.gov (accessed October 2012).

51. Jean Louis Touzet, 2013, 'La feuille de route d'Aqmi au Mali', *Libération*, 25 February, http://www.liberation.fr/monde/2013/02/25/la-feuille-de-route-d-aqmi-au-mali_884410 (accessed 2 May 2013).

52. Ibid.

53. Al-Jarrah, 2000, p. 58.

54. Halverson and Greenberg, 2017.

55. See Sara Cobb, 2013, *Speaking of Violence: the Politics and Poetics of Narrative Dynamics in Conflict Resolution*, Oxford: Oxford University Press, p. 138.

56. Abu Musab Abdel-Wadoud, 2007, 'AQLIM's Abdelouadoud issues statement warning collaborators, foreigners' [Where are the seekers of death?], *Al-Andalus Establishment for Media Production*, 10 May, http://www.opensource.gov (accessed 1 October 2012).

57. Al-Andalus Establishment for Media Production, 2011a.

58. Al-Andalus Establishment for Media Production, 2010a, 'AQLIM claims kidnapping of "five French nuclear experts" in Niger' ['Kidnapping five French nuclear experts in uranium mines in Niger'], *Al-Andalus Establishment for Media Production*, 22 September, http://www.opensource.gov (accessed 1 October 2012).

59. Shaykh Abu-Hayyan Assim, 2010, 'Audio message from AQLIM Shar'iah [sic] Commission Member' [Open letter to the people of the frontiers and their supporters by Shaykh Abu Hayyan Asim (May God protect him)], *Al-Andalus Establishment for Media Production*, 15 September, http://www.opensource.gov (accessed 1 October 2012).

60. Abou Abd Allah Ahmed, 2009, ['Interview with al-Andalus'], *Al-Andalus Establishment for Media Production*, 5 April, http://www.opensource.gov (accessed 1 October 2012).

61. Ibid.

62. Abdallah al-Jazairi, 2010, 'Forum participant warns against inciting Algerian tribes against AQLIM' [the scum apostates endeavor to propagate sedition between our kinfolk, the tribes and their sons of al-Qa'ida] [sic], *Al-Andalus Establishment for Media Production*, 5 April, http://www.opensource.gov (accessed 1 October 2012).

63. Ibid.

64. The United Nations Human Settlements Programme, 2012, *The State of Arab*

Cities 2012: Challenges of Urban Transition, http://www.unhabitat.org/pmss/ listItemDetails.aspx?publicationID=332 (accessed 1 May 2013).

65. Laurent de Saint Perier and Leïla Slimani, 2012, 'Le Qatar investi au Maghreb', *Jeune Afrique*, 21 February, http://www.jeuneafrique.com/142844/archives-thematique/le-qatar-investit-au-maghreb/ (accessed 27 February 2018).

66. Mag 14, 2013, 'La Tunisie cède 15% de Tunisiana à Qatar Telecom', *Mag 14*, 1 January, http://www.mag14.com/capital/62-entreprises/1366-la-tunisie-cede-15-de-tunisiana-a-qatar-telecom.html (accessed 27 February 2018).

67. See Yonah Alexander, 2013, Terrorism in North Africa and the Sahel in 2012, *The Potomac Institute*, February, http://www.potomacinstitute.org/images/ TerrorismNorthAfricaSahelGlobalReach.pdf (accessed 27 February 2018).

68. Al-Andalus Media Establishment for Media Production, 2011b, 'A new video message from Abu Mus'ab 'Abd al-Wadud ('Abd al-Malik Drūkdīl) the Amir of al-Qa'idah in the Islamic Maghreb: support for the free, descendants of Omar al-Mukhtar', http://jihadology.net/category/individuals/leaders/abu-mu%E1%B9%A3ab-abd-al-wadud-abd-al-malik-drukdil/page/2/ (accessed 21 June 2017).

Chapter 8

1. Taha Husayn, '*Adab al-thawra wa thawrat al-adab*', from the newspaper *Al-jumhuriya*, 23 October 1954. The article was reprinted in a special edition of *al-Hilal* in January 2015.

2. For an excellent discussion of this seminal work, see Mohammad Salama, 2018, *Islam and the Culture of Modernity: From the Monarchy to the Republic*, Cambridge: Cambridge University Press, pp. 79–83.

3. Ibid.

4. Ibid.

5. Ibid.

6. Ibid.

7. By 1962 there were fifteen theatres per 1 million habitants. This was still minor in comparison to Western Europe at this time where, according to Thorval, there existed on average 160 theatres for every 1 million habitants. See Yves Thoraval, 1975, *Regards Sur Le Cinéma Égyptien*, Beyrouth: Darl el-Machreq, p. 109. Also quoted in Nathaniel Greenberg, 2014, *The Aesthetic of Revolution in the Film and Literature of Naguib Mahfouz (1952–1967)*, Lanham: Lexington Books, p. xvii.

8. Ibid. pp. 107–8.

9. P. J. Vatikiotis, 1978, *Nasser and His Generation*, New York: St Martin's Press, p. 211.

10. Founded in 1935 by Talat Harb, Studio Misr quickly became the epicentre of film production in Egypt. Though Harb's dream of creating a Hollywood-like complex along the Avenue of the Pyramids in Giza never came to fruition, Studio Misr helped the Egyptian film industry grow to become the fourth largest in the world by 1952, with cotton alone exceeding the total revenue generated by film exports. In 1956, the Ministry of National Guidance was established to help channel funds towards production. This was followed by the Superior Council for the Protection of Arts and Letters as part of the newly established Ministry of Culture. Nasser nationalised Studio Misr in 1960. See Thoraval, 1975, pp. 107–8.

11. Husayn 2015.

12. Hani Mustafa, 2012, 'Shantytown Dogs', *Al-Ahram Weekly*, 15–29 July, http://weekly.ahram.org.eg/Archive/2012/1107/cu3.htm (accessed 23 October 2017).

13. Walter Armbrust, 1996, *Mass Culture and Modernism in Egypt*, New York: Cambridge University Press, p. 10.

14. See Fayrouz Karawya, 2010, 'Studio Misr to be auctioned off today', *Al-Masri al-Youm*, 18 July, http://www.egyptindependent.com/news/studio-misr-be-auctioned-today (accessed 10 May 2017).

15. Sarah Ibrahim, 2015, 'Egyptian cinema in crisis: the age of the low budget film', *Mada Masr*, 20 January, https://www.madamasr.com/en/2015/01/20/feature/culture/egyptian-cinema-in-crisis-the-age-of-the-low-budget-film/ (accessed 8 October 2017).

16. See Mohammed Baraka, 2008, 'Sinima fulus . . . Fulusna', *Youm 7*, 5 December, http://www.youm7.com/story/2008/12/5/%D8%A7%D9%84%D8%B3%D8%A8%D9%8A3%D9%89-%D8%B3%D9%8A%D9%86%D9%85%D8%A7-%D8%A7%D9%84%D9%81%D9%84%D9%88%D8%B3-%D9%81%D9%84%D9%88%D8%B3%D9%86%D8%A7-%D9%88%D8%A7%D8%AD%D9%86%D8%A7-%D8%B9%D8%A7%D8%B1%D9%81%D9%8A%D9%86-%D8%A7%D9%84%D9%86%D8%A7%D8%B3-%D8%B9%D8%A7%D9%8A%D8%B2%D9%87-%D8%A5%D9%8A%D9%87/54606 (accessed 18 December 2017).

17. Ibid.

18. Ibid. The phrase '*cigara wa kas*' (a smoke and a glass) stems from a song by Mohammed Abd al-Wahhab (d. 1991) of the same name.

19. Dr Naglaa Mahmoud Hussein, 2017, personal interview with the author, 1 October 2017, George Mason University.
20. See 'La l-il-fajr, la l-il-fajur', *RSSD*, 11 November 2015, https://www.youtube.com/watch?v=69fe23D-k_s (accessed 19 December 2017).
21. See Al-Ahram Online, 2011, 'Al-Haram Street makes Egyptian film history for first day release profits', *Al-Ahram Online*, 5 September, http://english.ahram.org.eg/NewsContent/5/32/20355/Arts--Culture/Film/El-Haram-Street-makes-Egyptian-film-history-for-bi.aspx (accessed 23 October 2017).
22. Baraka, 2008.
23. Amira Anwar, 2014, 'L-il-awwal marra fi tarikh al-sinima al-Misriya . . .', *Al-Ahram*, 14 August, http://www.ahram.org.eg/News/911/45/226482/%D8%B3%D9%8A%D9%86%D9%85%D8%A7/%D9%84%D8%A3%D9%88%D9%84-%D9%85%D8%B1%D8%A9-%D9%81%D9%8A-%D8%AA%D8%A7%D8%B1%D9%8A%D8%AE-%D8%A7%D9%84%D8%B3%D9%8A%D9%86%D9%85%D8%A7-%D8%A7%D9%84%D9%85%D8%B5%D8%B1%D9%8A%D8%A9%D9%81%D9%8A%D9%84%D9%85-%D9%8A%D8%AD%D8%B5%D8%AF%E2%80%8F-%E2%80%8F-%D9%85%D9%84%D8%A7%D9%8A.aspx (accessed 13 December 2017).
24. Mohamed Husayn, 2017, 'Rihlat Sa'ud al-Batal', *Huffpost Arabic*, 1 December, http://www.huffpostarabi.com/2017/11/28/story_n_18671064.html (accessed February).
25. See Nathaniel Greenberg, 2014, *The Aesthetic of Revolution in the Film and Literature of Naguib Mahfouz (1952–1967)*, Lanham: Lexington Books, pp. 4–9.
26. Al-Ahram Online, 2011.
27. Mustafa, 2012.
28. Mohamad Elmeshad, 2011, 'Crime on the rise, but are police to blame?', *Al-Masri al-Youm*, 23 May, http://www.egyptindependent.com/crime-rise-are-police-blame/ (accessed 23 October 2017).
29. Ahmed Morsy, 2013, 'Rough justice', *Al-Ahram Weekly*, 21 March, http://weekly.ahram.org.eg/News/1946.aspx (accessed 23 October 2017).
30. Mayy El Sheikh and David Kirkpatrick, 2013, 'Rise in sexual assaults in Egypt sets off clash over blame', *The New York Times*, 25 March, http://www.nytimes.com/pages/world/index.html (accessed 25 March 2013).
31. Adel Iskander pointed this out in a lucid piece for the *Huffington Post* in 2011. See Iskander, 'The Baltageya: Egypt's counterrevolution', www.huffington-

post.com/adel-iskandar/the-baltageya-egypts-coun_b_862267.html (accessed 24 October 2017).

32. Jacques Rancière, 2004, *The Politics of Aesthetics: the Distribution of the Sensible*, London: Continuum, p. 24.

33. Ibid. p. 24.

34. Wilson Chacko Jacob, 2007, 'Eventful transformations: al-futuwwa between history and the everyday', *Comparative Studies in Society and History* 49.3.

35. Elliot Colla, 2005, 'Anxious advocacy: the novel, the law, and the extrajudicial appeals in Egypt', *Public Culture* 17.3, 436.

36. See P. J. Vatikiotis, 1971, 'The corruption of futūwa: a consideration of despair in Nagib Maḥūẓ's Awlād Ḥāratinā', *Middle Eastern Studies* 7.2, 169–84.

37. Belal Fadl, 2011, 'Al-Baltagiya 7 najum', *Facebook*, 11 October, https://www.facebook.com/note.php?note_id=254919557888186 (accessed 14 December 2017).

38. Néstor García Canclini, 1995, *Hybrid Cultures: Strategies for Entering and Leaving Modernity*, Minneapolis: University of Minnesota Press, p. 188.

39. Ibid. p. 188.

40. Slim Amamou, 2012, 'NoMemorySpace', *Wordpress*, 31 May, https://nomemoryspace.wordpress.com/.

41. Lauren Cochrane, 2010, 'Romain Gavras: Born Free director is no stranger to Stress', *The Guardian*, 24 September, https://www.theguardian.com/music/2010/sep/25/romain-gavras-born-free (accessed 3 February 2018).

42. Michel Ciment, 1983, *Kubrick*, New York: Holt, Reinhart and Winston, p. 163.

43. Ahmed Khaled Tawfiq, 2008, *Yutubiya*, Doha: Bloomsbury Qatar Foundation, p. 5.

44. Ibid. p. 116.

45. Ibid. p. 118.

46. Ibid. p. 118.

47. On the 'project of succession' (*mashruʿ al-tawrith*), see Chapter 2.

48. Ahmed Khaled Tawfiq, 2012, *al-Sinja*, Cairo: Dar al-Shorouk, p. 203.

49. Ibid. p. 24.

50. Ibid. p. 224.

51. *Al-Sija* is an ancient game similar to checkers. It is traditionally played with stones.

52. Tawfiq, 2012, p. 209.

53. Ibid. p. 211.
54. Ibid. p. 211.
55. Ibid. p. 239.
56. See Abdallah Shawikh, 2017, 'Ahmed Khaled', *Emarat al-youm*, 24 March, http://www.emaratalyoum.com/life/culture/2017-03-25-1.981006 (accessed 27 February 2018).
57. Amir Zaky, 2012, 'When Dracula speaks', *Egypt Independent*, 11 August, http://www.egyptindependent.com/news/when-dracula-speaks-arabic (accessed 27 February 2018).
58. See Mohamed Abd al-Rahman, 2016, '*Ta'rif 'ala qawa'im . . .*', *Youm 7*, 31 October, http://www.youm7.com/story/2016/10/31/%D8%AA%D8%B9%D8%B1%D9%81-%D8%B9%D9%84%D9%89-%D9%82%D9%88%D8%A7%D8%A6%D9%85-%D8%A7%D9%84%D8%A3%D9%83%D8%AB%D8%B1-%D9%85%D8%A8%D9%8A%D8%B9%D8%A7-%D8%A8%D8%A7%D9%84%D9%85%D9%83%D8%AA%D8%A8%D8%A7%D8%AA-%D8%A7%D9%84%D9%85%D8%B5%D8%B1%D9%8A%D8%A9-%D9%84%D8%B4%D9%87%D8%B1-%D8%A3%D9%83%D8%AA%D9%88%D8%A8%D8%B1/2943468 (accessed 27 March 2017).
59. Akhbar al-awan, 2014, 'Al-Katib Ahmed Khaled Tawfiq', *Akhbar al-awan*, 2 February, http://news.asu.edu.eg/uninews.php?action=show&nid=5190&type=292 (accessed 28 March 2017).
60. UNICEF Statistics, 2017, https://www.unicef.org/infobycountry/egypt_statistics.html#117 (accessed 28 March 2017).
61. CIA, 2017, *CIA World Factbook*, https://www.cia.gov/library/publications/the-world-factbook/fields/2103.html (accessed 28 March 2017).
62. Mohamed al-Buali, 2015, 'Kayf ghayr al-shabab suq al-kitab fi Misr', *Goethe Institute*, 20 November, https://www.goethe.de/ins/eg/ar/kul/mag/20717931.html (accessed 27 March 2017).
63. CIA, 2017.
64. Ahmed Khaled Tawfiq, 2017, *Al-Ghuz wara' al-sotur*, Cairo: Dar al-Shorouk, p. 114.
65. See Israel Gershoni and James P. Jankowski, 1986, *Egypt, Islam, and the Arabs: The Search for Egyptian Nationhood, 1900–1930*, New York: Oxford University Press, p. 109.
66. See G. Willow Wilson and Rebecca Hankins, 'Islam Sci-Fi interview with G. Willow Wilson', *Islamscifi*, 25 March 2015, http://www.islamscifi.com/

islam-sci-fi-interview-of-g-willow-wilson-part-i/. There exists a veritable deluge of blogs devoted to historicising Arabic science fiction.

67. See Yusuf Sharuri, 2000, *Al-Khayal al-'ilmi fi al-adab al-'arabi al-mu'asir: hatta nihayat al-qarn al-'ishrin*, Cairo: al-Haya al-Misriya al-'Amma lil-Kitab; or 2006, 'Adab al-khayal al-'ilmi al-'arabi', *Al-Ghrad*, 26 February, http:// www.alghad.com/articles/772434-%D8%A3%D8%AF%D8%A8-%D8% A7%D9%84%D8%AE%D9%8A%D8%A7%D9%84-%D8%A7%D9% 84%D8%B9%D9%84%D9%85%D9%8A-%D8%A7%D9%84%D8% B9%D8%B1%D8%A8%D9%8A-%D9%8A%D8%AD%D8%AA%D8% A7%D8%AC-%D8%AA%D9%82%D8%AF%D9%85%D8%A7-%D8% AA%D9%82%D9%86%D9%8A%D8%A7-%D9%8A%D9%86%D8% AA%D8%AC-%D9%84%D8%BA%D8%A9-%D9%85%D8%BA%D8% A7%D9%8A%D8%B1%D8%A9.

68. Najib Mahfuz, 2006, *Awlad haratina*, Al-Qahira: Dar al-Shorouk, p. 523.

69. Ibid. p. 580.

70. Nihad Sharif, 1972, *Qahir al-zaman*, Cairo: Dar al-Hilal, p. 41.

71. As discussed by Ian Campbell, influence of Wells' *The Time Machine* can also be read in Tawfiq's *Utopia*. See Ian Campbell, 2015, 'Prefiguring Egypt's Arab Spring: allegory and allusion in Ahmad Khālid Tawfīq's Utopia', *Science-fiction Studies* 42, 544–5.

72. Ada Barbaro, 2013, *La Fantascienza Nella Letteratura Araba*, Roma: Carocci, p. 152.

73. Fredric Jameson, 2007, *Archaeologies of the Future: The Desire Called Utopia and Other Science Fictions*, London: Verso, p. 95.

74. Ibid. p. 95.

75. Ghali Shukri, 1981, *Egypt, Portrait of a President, 1971–1981: the Counter-revolution in Egypt, Sadat's Road to Jerusalem*, London: Zed Press, p. 198.

76. Abd al-Latif al-Soltani, 1974, *Al-Mazdakiya hiya 'asal al-ishtirakiya*, Casablanca: Aljaza'ir, Jami'at al-haquq mahfuza l-il-mu'lif, p. 11.

77. On Fourth Generation Warfare in Egyptian state discourse, see Chapter 4.

78. Ahmed Khaled Tawfiq, 2015, *Mithl Ikarus: Riwaya*, Cairo: Dar al-Shorouk, p. 97.

79. Ibid. p. 26.

80. Ibid. p. 97.

81. Ibid. p. 174.

82. Al-Ahram, 2011, 'Iqalat al-hukuma', *Al-Ahram* (Cairo), 29 January, p. 1.

83. Tawfiq, 2015, p. 330.

84. Ibid. p. 97.

85. Ahmed Khaled Tawfiq, 2011, 'Inihum ya'kulun al-Kintaki', Aktowfik Blog, 3 February, http://aktowfik.blogspot.com/2011/02/blog-post_05.html (accessed 28 February 2018).

86. Tawfiq, 2012, p. 10.

87. Ibid. p. 9.

88. The phrase 'killing to have hunted' stems from Jaroslav Stetkevych's observation of the pre-Islamic poet Imru al-Qays whose pioneering *tardiya*, or hunt poem, would become a stable of Arabic literary expression for centuries to come. See Jaroslav Stetkevych, 2016, *The Hunt in Arabic Poetry*, Notre Dame: Notre Dame University Press, p. 1.

89. Tawfiq, 2015, p. 9.

90. Kelmetna, 2011, 'Mahmoud 'Athman . . .', *Kelmetna*, 12 March, http://www.masress.com/kelmetna/10505 (accessed 28 February 2018).

Bibliography

Abbas, Wael, 2007, 'Help our fight for real democracy', *The Washington Post*, 27 May, http://www.washingtonpost.com/wp-dyn/content/article/2007/05/25/AR2007052502024_pf.html (accessed 13 July 2017).

Abd al-Hadi, Muhammad, 2011, 'Ihtijajat wa idtirabat was'a fi Libnan', *Al-Ahram* (Cairo), 26 January.

Abd al-Rahman, Abu, 2010, 'Letter to Salah Abu Muhammad', *Dni.gov*, https://www.dni.gov/index.php (accessed 18 January 2018).

Abd al-Rahman, Mohamed, 2016 '*Ta'rraf* 'ala quwa'im . . .', *Youm 7*, 31 October, http://www.youm7.com/story/2016/10/31/%D8%AA%D8%B9%D8%B1%D9%81-%D8%B9%D9%84%D9%89-%D9%82%D9%88%D8%A7%D8%A6%D9%85-%D8%A7%D9%84%D8%A3%D9%83%D8%AB%D8%B1-%D9%85%D8%A3%D9%8A%D8%B9%D8%A7-%D8%A8%D8%A7%D9%84%D9%85%D9%83%D8%AA%D8%A8%D8%A7%D8%AA-%D8%A7%D9%84%D9%85%D8%B5%D8%B1%D9%8A%D8%A9-%D9%84%D8%B4%D9%87%D8%B1-%D8%A3%D9%83%D8%AA%D9%88%D8%A8%D8%B1/2943468 (accessed 27 March 2017).

Abd Al-Rahman, Walid, 2012, 'Ramzi: Akthar min 100 alf qibti Misri taqdimu bi-il-talbat hijra', *Al-Arabiya*, 10 September, https://www.alarabiya.net/articles/2012/09/10/237094.html (accessed 30 October 2017).

Abd al-Rauf, Mohammed, 2011, 'Al-i'lam al-hukumi al-Misri yusara' fi-al-lihaq bi-qitar al-thawra', *Al-Sharq al-awsat*, 13 February, http://archive.aawsat.com/details.asp?section=4&article=607997&issueno=11765#.V62PPqJp5WU (accessed 20 October 2017).

Abd al-Shokur, Muhammad, 2012 'Video: Al-Nas tadhi' laqatat min al-film al-musi' l-il-rasul', *Al-Wafd*, 9 September, https://alwafd.org/%D8%AF%D9%86%D9%8A%D8%A7-%D9%88%D8%AF%D9%8A%D9%86/262446-%D9%81%D9%8A%D8%AF%D9%8A%D9%88-%D8%A7%D9%84%D9%86%

D8%A7%D8%B3-%D8%AA%D8%B0%D9%8A%D8%B9-%D9%84%
D9%82%D8%B7%D8%A7%D8%AA-%D9%85%D9%86-%D8%A7%
D9%84%D9%81%D9%8A%D9%84%D9%85-%D8%A7%D9%84%D9%
85%D8%B3%D9%89%D8%A1-%D9%84%D9%84%D8%B1%D8%B3%
D9%88%D9%84 (accessed 30 October 2017).

Abdelkader, Engy, 'The anatomy of a terrorist designation: the Muslim Brotherhood and international terrorism', Penn Law Global Affairs Blog, https://www.law.upenn.edu/live/news/6858-the-anatomy-of-a-terrorist-designation-the-muslim/news/international-blog.php (accessed 1 May 2018).

Abdel-Wadoud, Abu Musab, 2007, 'AQLIM's Abdelouadoud issues statement warning collaborators, foreigners' [Where are the seekers of death?], *Al-Andalus Establishment for Media Production*, 10 May, http://www.opensource.gov (accessed October 2012).

Abduh, Ibrahim, 1964, *Jaridat al-Ahram: tarikh wa al-fann 1875–1964*, al-Qahira: Mu'assasat Sijill al-'Arab.

Abdulla, Rasha, 2014, 'Egypt's media in the midst of revolution', *The Carnegie Endowment for International Peace*, 16 July, http://carnegieendowment.org/2014/07/16/egypt-s-media-in-midst-of-revolution-pub-56164 (accessed 7 June 2017).

Abu Muhammad, Salah, 2011, 'AQLIM says *Al-Hayat* interview with spokesman "fake"', *Al-Andalus Establishment for Media Production*, http://www.opensource.gov (accessed 20 May 2013).

Adunis, 2011, 'Ramad al-Bu'azizi', *Al-Arabiya*, 28 April, https://www.alarabiya.net/views/2011/04/28/147037.html (accessed 24 February 2018).

Afify, Heba, 2014, 'Egyptian media isn't taking prisoners, State's line is only line', *Mada Masr*, 27 October 2014, http://www.madamasr.com/en/2014/10/27/feature/politics/egyptian-media-isnt-taking-prisoners-states-line-is-only-line/ (accessed 23 May 2017).

Al-Ahram, 2011, 'Al-alaf yasharikun fi mudhahirat selmiya bi-al-Qahira wa al-muhafizat', *Al-Ahram* (Cairo), 26 January.

—, 2011, 'Rashid: la narghabu fi ittifaq tijara hurra ma'a Amrika', *Al-Ahram* (Cairo), 26 January, p. 1.

—, 2011, 'Wafaa 4 wa asaba 118 . . .', *Al-Ahram* (Cairo), 27 January.

—, 2011, 'Khatar al-tajwal fi al-Qahira . . .', *Al-Ahram* (Cairo), 28 January, p. 1.

—, 2011, 'Burqiyat diblumasiya hasal 'alayha WikiLeaks', *Al-Ahram* (Cairo), 29 January, p. 4.

—, 2011, 'Hazr al-tajawwul fi al-Qahira', *Al-Ahram* (Cairo), 29 January, p. 1.

—, 2011, 'Iqalat al-hakuma', *Al-Ahram* (Cairo), 29 January.

—, 2011, 'Ijtima' l-il-ra'is Mubarak ma'a al-qiyadat al-'askariya', *Al-Ahram* (Cairo), 31 January. p. 1.

—, 2011, 'Hukuma jadida bi-la 'rijal a'amal', *Al-Ahram* (Cairo), 1 February.

—, 2011, 'Al-Malayin yukhrujun . . .', *Al-Ahram* (Cairo), 3 February.

—, 2011, 'Suleiman: 'al-huwwar' aw 'al-inqilab', *Al-Ahram* (Cairo), 9 February, p. 1.

—, 2011, 'Al-sha'b asqata al-nizam', *Al-Ahram* (Cairo), 12 February, p. 1.

Al-Ahram Online, 2011, 'Al-Haram Street makes Egyptian film history for first day release profits', *Al-Ahram Online*, 5 September, http://english.ahram.org. eg/NewsContent/5/32/20355/Arts--Culture/Film/El-Haram-Street-makes-Egyptian-film-history-for-bi.aspx (accessed 23 October 2017).

—, 2015, 'Trials of producer El-Sobki, actress Entsar over "immoral actions" are adjourned', *Ahram Online*, 15 December, http://english.ahram.org.eg/ NewsContent/5/159/173465/Arts--Culture/Entertainment/Trials-of-pro ducer-ElSobki,-actress-Entsar-over-im.aspx (accessed 19 December 2017).

Akhbar al-awan, 2014, 'Al-Katib Ahmed Khaled Tawfiq', *Akhbar al-awan*, 2 February, http://news.asu.edu.eg/uninews.php?action=show&nid=5190&type=292 (accessed 28 March 2017).

Alexander, Yonah, 2013, 'Terrorism in North Africa and the Sahel in 2012', *The Potomac Institute*, February, http://www.potomacinstitute.org/images/ TerrorismNorthAfricaSahelGlobalReach.pdf (accessed 27 February 2018).

Amami, Azyz, 2012, 'My Palestine', *Al-Arabiya English*, 22 November, https:// english.alarabiya.net/authors/Azyz-Amami.html (accessed 20 February 2013).

Amamou, Slim, 2012, 'NoMemorySpace', *Wordpress*, 31 May, https://nomemory space.wordpress.com/.

Ammar, Amr, 2014, *Al-Intilal al-madani: asrar 25 Yanayir wa-al-Marinz al-Amiriki: hurub al-jil al-rabi' min al-thawarat al-mulawwana ila al-Rabi' al-Arabi*, Cairo: Tawzi' al-Majmu'a al-dawliya l-il-nashr wa al-tawzi'.

Al-Andalus Establishment for Media Production, 2010a, 'AQLIM claims kid- napping of "five French nuclear experts" in Niger' ['Kidnapping five French nuclear experts in uranium mines in Niger'], *Al-Andalus Establishment for Media Production*, 22 September, http://www.opensource.gov (accessed 1 October 2012).

—, 2010b, 'Tubdi' wa taqdim', *Archive.org*, https://archive.org/details/Elhaq-Be-Al-Qafelah (accessed 27 February 2018).

—, 2011a, 'Al-Andalus releases audio by AQLIM leader on recent unrest in Tunisia' ['Help and support for our people's uprising in Tunisia'], *Al-Andalus Establishment for Media Production*, 13 January, http://www.opensource.gov (accessed October 2012).

—, 2011b, 'A new video message from Abu Mus'ab 'Abd al-Wadud ('Abd al-Malik Drukdil) the Amir of al-Qa'idah in the Islamic Maghreb: support for the free, descendants of Omar al-Mukhtar', http://jihadology.net/category/individuals/ leaders/abu-mu%E1%B9%A3ab-abd-al-wadud-abd-al-malik-drukdil/page/2/ (accessed 21 June 2017).

—, 2012a, ['In the fields and squares of liberation . . .'], *Al-Andalus Establishment for Media Production*, 22 May, http://opensource.gov (accessed October 2012).

—, 2012b, ['Our Muslim brothers in Tunisia'], *Al-Andalus Establishment for Media Production*, 22 October, http://opensource.gov (accessed October 2012).

—, 2014, *al-Jaza'ir wa al-nafaq al-muzlim*, https://www.youtube.com/watch?v= 27iTIWAin78&t=21s (accessed 15 June 2017).

Anderson, Benedict, 1983, *Imagined Communities*, London: Verso.

Anishchuk, Alexei, 2014, 'Russia to boost trade with Egypt after Western food ban', *Reuters*, 12 August, https://www.reuters.com/article/us-ukraine-crisis-russia-egypt/russia-to-boost-trade-with-egypt-after-western-food-ban-idUSK BN0GC1D820140812 (accessed 25 February 2018).

al-Annabi, Abu Obayda Yusuf, 2011, ['An interview with Abu Ubaydah Yusuf al-Annabi'], *Al-Andalus Establishment for Media Production*, 7 July, http://www. opensource.gov (accessed October 2012).

al-Ansari, Amru Ahmed, 2014, *Jumhuriyat al-Ultras: 7 sanwat ashghal shaqqa*, Cairo: Nahdha Misr lil-Nashr.

Anwar, Amira, 2014, 'L-il awwal marra fi tarikh al-sinima al-Misriya . . .', *Al-Ahram*, 14 August, http://www.ahram.org.eg/News/911/45/226482/%D8%B3% D9%8A%D9%86%D9%85%D8%A7/%D9%84%D8%A3%D9%88%D9% 84-%D9%85%D8%B1%D8%A9-%D9%81%D9%8A-%D8%AA%D8% A7%D8%B1%D9%8A%D8%AE-%D8%A7%D9%84%D8%B3%D9% 8A%D9%86%D9%85%D8%A7-%D8%A7%D9%84%D9%85%D8%B5% D8%B1%D9%8A%D8%A9%D9%81%D9%8A%D9%84%D9%85-% D9%8A%D8%AD%D8%B5%D8%AF%E2%80%8F-%E2%80%8F-% D9%85%D9%84%D8%A7%D9%8A.aspx (accessed 13 December 2017).

Applebaum, Anne, 2017, 'If Russia can create fake "Black Lives Matter" accounts, who will be next?', *The Washington Post*, 16 October, https://www.washington

post.com/opinions/global-opinions/if-russia-can-create-fake-black-lives-matter-accounts-who-will-next/2017/10/15/ffb2e01e-af79-11e7-be94-fabb0f1e9ffb_story.html?utm_term=.11cb3d91ba11 (accessed 16 October 2017).

AQIM Media Committee, 2009, 'AQLIM: "Declaration of al-Andalus Establishment for Media Production"' [Good news: Declaration of al-Andalus Establishment for Media Production], *Al-Andalus Establishment for Media Production*, 4 October, http://www.opensource.gov (accessed October 2012).

Aqlame, 2014, 'I'lan madinat Ghao imara tabi'a l-il-da'ish', *Aqlame.com*, http://www.aqlame.com/article19431.html (accessed 9 January 2018).

al-Aqqad, Abbas Mahmud, 1952, 'Adab al-nahdha al-jadida', *Al-Hilal*, 12.60.

Al-Arabiya, 2011, "'Amr Musa yurrahibu bi-thawrat al-Bayda' . . .', *Al-Arabiya*, 11 February, http://www.alarabiya.net/articles/2011/02/11/137244.html (accessed 1 February 2017).

—, 2013, 'Al-Ghannouchi: Bel'id laysa al-Bu'azizi wa ana laysa Ben 'Ali', *Al-Arabiya*, 10 February, http://www.alarabiya.net/articles/2013/02/10/265374.html (accessed 3 October 2017).

—, 2013, ''Abu 'Ayadh al-Tunisi: ikhtitaf raha'in al-Jaza'ir irhab "Mahmud"', *al-Arabiya*, 12 February, https://www.alarabiya.net/ar/arab-and-world/2013/02/12/%D8%A3%D8%A8%D9%88-%D8%B9%D9%8A%D8%A7%D8%B6-%D8%A7%D9%84%D8%AA%D9%88%D9%86%D8%B3%D9%8A-%D8%A7%D8%AE%D8%AA%D8%B7%D8%A7%D9%81-%D8%B1%D9%87%D8%A7%D8%A6%D9%86-%D8%A7%D9%84%D8%AC%D8%B2%D8%A7%D8%A6%D8%B1-%D8%A5%D8%B1%D9%87%D8%A7%D8%A8-%D9%85%D8%AD%D9%85%D9%88%D8%AF--1418.html (accessed 28 February 2018).

Armbrust, Walter, 1996, *Mass Culture and Modernism in Egypt*, New York: Cambridge University Press.

—, 2013, 'The trickster in Egypt's January 25 Revolution', *Comparative Studies in Society and History* 55.4, 834–64.

As'ad, Mohamed, 2016, 'Hay'at jadida l-il-sihafa wa al-i'lam', *Youm 7*, 19 November, http://www.youm7.com/story/2016/11/19/3-%D9%87%D9%8A%D8%A6%D8%A7%D8%AA-%D8%AC%D8%AF%D9%8A%D8%AF%D8%A9-%D9%84%D9%84%D8%B5%D8%AD%D8%A7%D9%81%D8%A9-%D9%88%D8%A7%D9%84%D8%A5%D8%B9%D9%84%D8%A7%D9%85-%D9%86%D9%86%D8%B4%D8%B1-%D9%85%D9%84%D8%A7%D9%85%D8%AD-%D8%AA%D8%B4%D9%83%D9%8A%D9%84-%D8%A7%D9%84%D9%85%D8%AC%D9%84%D8%

B3-%D8%A7%D9%84%D8%A3%D8%B9%D9%84%D9%89/2974004
(accessed 20 July 2017).

Asch, Solomon E., 1952, *Social Psychology*, Englewood Cliffs: Prentice Hall Inc.

—, 1987, *Social Psychology*, Oxford: Oxford University Press.

al-Ashul, Ismail, 2014, 'Hamdi Qandil', *Al-Shorouk*, February, http://www.shorouk
news.com/news/view.aspx?cdate=04022014&id=67fe45ae-652c-42fd-afa8-
dba8319af934 (accessed 25 February 2018).

Asim, Shaykh Abu-Hayyan, 2010, 'Audio message from AQLIM Shar'iah [sic]
Commission Member' [Open letter to the people of the frontiers and their
supporters by Shaykh Abu Hayyan Asim (May God protect him)], *Al-Andalus
Establishment for Media Production*, 15 September, http://www.opensource.gov
(accessed 10 May 2013).

Askari, Rasad, 2016, 'Limadha tudariba Rusya junudaha 'ala harub al-Sahara' fi
Misr', *Sputnik*, 12 October, https://arabic.sputniknews.com/military/201610
121020433997-%D8%B1%D9%88%D8%B3%D9%8A%D8%A7-%D9%
85%D8%B5%D8%B1-%D8%A3%D9%85%D8%B1%D9%8A%D9%
83%D8%A7-%D8%A7%D9%84%D8%AC%D9%8A%D8%B4-%D8%
A7%D9%84%D9%85%D8%B5%D8%B1%D9%8A-%D9%84%D9%8A%
D8%A8%D9%8A%D8%A7/ (accessed 23 February 2018).

Assange, Julian, 2016, *When Google met WikiLeaks*, London: OR Books.

al-Aswany, Alaa, 2013, *The Automobile Club*, Cairo: Dar al-Shorouk.

Attal, Sylvain, 2011, 'Tahar Ben Jelloun: Écrivain', *France 24*, 14 July, http://www.
france24.com/fr/20110713-tahar-ben-jelloun-etincelle-par-le-feu-revolution-
arabe-revolte-vendeur-tunisie-tunis-bouazizi (accessed 24 February 2018).

Austin, J. L., 2011, *How to do Things with Words: the William James Lectures Delivered
at Harvard University in 1955*, Oxford: Clarendon Press.

Ayalon, Ami, 2016, *The Arabic Print Revolution: Cultural Production and Mass
Readership*, Cambridge: Cambridge University Press.

Ayari, Yassine, 2011, 'Caid Essebsi, Marzouki, Rached Ghannouchi, Nejib Chebbi
et la malaise de Yassine Ayari', *Mel7it* blog, 28 April, http://mel7it3.blogspot.
com/2011/04/caid-essebsi-marzouki-rached-ghannouchi.html (accessed 23 July
2017).

Baker, Aryn, 2014, 'Why Iran believes the militant group ISIS is an American plot',
Time, 19 July, http://time.com/2992269/isis-is-an-american-plot-says-iran/
(accessed 24 February 2018).

Bakri, Mustafa, 2015, *Al-Sisi: al-tariq ila bina' al-dawla*, Cairo: Dar al-Masriya
al-Libnaniya.

Baraka, Mohammed, 2008, 'Sinima fulus ... Fulusna', *Youm 7*, 5 December, http://www.youm7.com/story/2008/12/5/%D8%A7%D9%84%D8%B3% D8%A8%D9%83%D9%89-%D8%B3%D9%8A%D9%86%D9%85%D8% A7-%D8%A7%D9%84%D9%81%D9%84%D9%88%D8%B3-%D9% 81%D9%84%D9%88%D8%B3%D9%86%D8%A7-%D9%88%D8%A7% D8%AD%D9%86%D8%A7-%D8%B9%D8%A7%D8%B1%D9%81% D9%8A%D9%86-%D8%A7%D9%84%D9%86%D8%A7%D8%B3-% D8%B9%D8%A7%D9%8A%D8%B2%D9%87-%D8%A5%D9%8A% D9%87/54606 (accessed 18 December 2017).

Barbaro, Ada, 2013, *La Fantascienza Nella Letteratura Araba*, Roma: Carocci.

BBC, 2011, 'Tunisia suicide protester Mohammed Bouazizi dies', *BBC*, 5 January, http://www.bbc.com/news/world-africa-12120228 (accessed 16 March 2017).

—, 2017, 'Cambridge scientists consider fake news "vaccine"', *BBC*, 23 January, http://www.bbc.com/news/uk-38714404 (accessed 30 June 2017).

Ben Gharbia, Sami, 2003, *Borj Erroumi XL: voyage dans un monde hostile*, https://ifikra.files.wordpress.com/2009/12/borjerroumixl.pdf (accessed 24 February 2018).

—, 2008, 'Tunisia seems to have blocked access to Facebook', *Global Voices*, 18 August https://advox.globalvoices.org/2008/08/18/tunisia-seems-to-have-bloc ked-access-to-facebook/.

—, 2010a, 'Anti-censorship movement in Tunisia: creativity, courage and hope!', *Global Voices*, 27 May, https://advox.globalvoices.org/2010/05/27/anti-censor ship-movement-in-tunisia-creativity-courage-and-hope/print/.

—, 2010b, 'The Internet freedom fallacy and the Arab digital activism', *Nawaat*, 17 September, https://nawaat.org/portail/2010/09/17/the-internet-freedom- fallacy-and-the-arab-digital-activism/ (accessed 18 February 2018).

—, 2012, 'Curating reality: new tools for investigative journalism', Sandberg @ mediafonds conference, Amsterdam, Netherlands, 9 February, https://vimeo. com/36764899 (accessed 11 February 2018).

—, 2017, 'Shahadat Sami Ben Gharbia', *Instance Vérité & Dignite Commission, Tunis*, 11 March, https://www.youtube.com/watch?v=ZOC3Rp9Zvks (accessed 1 September 2017).

Benjamin, Walter, 1955 *Illuminations*, New York: Schocken Books.

Ben Mhenni, Lina, 2008, 'Tunisia: National day for freedom of blogging on November 4', *Global Voices*, 10 October, https://globalvoices.org/2008/10/10/ tunisia-national-day-for-freedom-of-blogging-on-november-4/ (accessed 26 February 2018).

—, 2010, 'Tunisia: Unemployed man's suicide attempt sparks riots', *Global Voices*, 23 December, https://globalvoices.org/2010/12/23/tunisia-unemployed-mans-suicide-attempt-sparks-riots/ (accessed 1 February 2018).

—, 2011, *Tunisian Girl/Bunaiyat Tunissiya: Blogueuse pour un printemps arabe*, Montpellier: Indigène.

—, 2017, '13 Mars', *Linkedin*, 17 March, https://fr.linkedin.com/pulse/13-mars-lina-ben-mhenni (accessed 15 February 2018).

Berlin, Isaiah, 1958, 'The origins of Israel', in Walter Z. Laqueur (ed.), *The Middle East in Transition*, New York: Praeger.

Bickert, Monika, 2016, 'Internet security and privacy in the age of the Islamic State: The view from Facebook', *Washington Institute for Near East Policy*, 26 February, http://www.washingtoninstitute.org/policy-analysis/view/internet-security-and-privacy-in-the-age-of-the-islamic-state (accessed 2 June 2016).

Bidoun, 2011, 'Enough is not enough: Abdel Halim Qandil', *Bidoun #25*, http://bidoun.org/articles/enough-is-not-enough (accessed 1 June 2016).

Blaise, Lilia, 2014, 'De l'ATI à l'ATT: Quel avenir pour l'internet en Tunisie?', *Huffpost Maghreb*, 18 March, http://www.huffpostmaghreb.com/2014/03/18/internet-tunisie_n_4985350.html (accessed 23 February 2018).

Bok, Sissela, 1989, *Lying: Moral Choice in Public and Private Life*, New York: Vintage.

Booth, Robert, 2010, 'Cables claim al-Jazeera changed coverage to suit Qatari foreign policy', *The Guardian*, 5 December, https://www.theguardian.com/world/2010/dec/05/wikileaks-cables-al-jazeera-qatari-foreign-policy (accessed 30 May 2018).

Borges, Jorge Luis, 1977, *El libro de arena*, Buenos Aires: Biblioteca Borges.

al-Bouti, Mounira, 2016, 'La mère de Nadhir Ktari, disparu en Libye: "Notre visite a été fructueuse"', *Huffpost Maghreb*, 5 September, http://www.huffpostmaghreb.com/2016/09/05/sofiane-chourabi-nehdir-_n_11864758.html (accessed 30 January 2018).

Bradley, Matt, 2012, 'Missions stormed in Libya, Egypt', *WSJ*, 12 September, https://www.wsj.com/articles/SB10000872396390444017504577645681057498266 (accessed 30 October 2017).

Brand, Laurie, 2014, *Official Stories: Politics and National Narratives in Egypt and Algeria*, Palo Alto: Stanford University Press.

Bressan, Yannick, 2012, *Du principe d'adhésion au théâtre: Approche historique et phénoménologique*, Paris: L'Harmattan.

—, 2018, *Radicalisation, renseignement et individus toxiques: mieux comprendre les processus de manipulation mentale*, Paris: VA éditions.

Briant, Emma L., 2015, *Propaganda and Counter-Terrorism: Strategies for Global Change*, Manchester: Manchester University Press.

Bryanski, Gleb, 2012, 'Youtube under threat in Russia over prophet film', *Reuters*, 18 September, http://www.reuters.com/article/us-protest-russia/youtube-under-threat-in-russia-over-prophet-film-idUSBRE88H16L20120918 (accessed 1 November 2017).

al-Buali, Mohamed, 2015, 'Kayf ghayr al-shabab suq al-kitab fi Misr', *Goethe Institute*, 20 November, https://www.goethe.de/ins/eg/ar/kul/mag/20717931.html (accessed 27 March 2017).

Bughlab, Muhammad, 2012, 'Al-Shaykh "Khamis al-Majri" fi hiwar khas ma'a al-Tunisiya', *Al-Tunisiya*, 18 April, http://www.attounissia.com.tn/details_article.php?t=37&a=56222 (accessed 20 April 2012).

Burgat, François, and William Dowell, 1997, *The Islamic Movement in North Africa*, Austin: Center for Middle Eastern Studies, University of Texas at Austin.

Business News, 2018, 'Mohamed Ghariani: Des blogueurs et des politiciens ont été formés pour renverser l'ancien régime!', *Business News*, 5 January, http://www.businessnews.com.tn/comment/874759 (accessed 15 February 2018).

Cahen, Claude, 1959, *Mouvements populaires et autonomisme urbain dans l'asie Musulmane du Moyen Âge*, Leiden: E. J. Brill.

Campbell, Ian, 2015, 'Prefiguring Egypt's Arab Spring: allegory and allusion in Ahmad Khalid Tawfiq's Utopia', *Science-fiction Studies* 42, 544–5.

Canclini, Néstor García, 1995, *Hybrid Cultures: Strategies for Entering and Leaving Modernity*, Minneapolis: University of Minnesota Press.

Cherribi, Sam, 2017, *Fridays of Rage: Al-Jazeera, the Arab Spring, and Political Islam*, Oxford: Oxford University Press.

Chick, Kristen, 2011, 'How revolt sparked to life in Tunisia', *Christian Science Monitor*, 23 January, https://www.csmonitor.com/World/Middle-East/2011/0123/How-revolt-sparked-to-life-in-Tunisia (accessed 25 July 2017).

CIA, 2017, *CIA World Factbook*, https://www.cia.gov/library/publications/the-world-factbook/fields/2103.html (accessed 28 March 2017).

Ciment, Michel, 1983, *Kubrick*, New York: Holt, Reinhart and Winston. [Courtesy Willis Konick.]

Cobb, Sara B, 2013, *Speaking of Violence: the Politics and Poetics of Narrative Dynamics in Conflict Resolution*, Oxford: Oxford University Press.

Cochrane, Lauren, 2010, 'Romain Gavras: Born Free director is no stranger to Stress', *The Guardian*, 24 September, https://www.theguardian.com/music/2010/sep/25/romain-gavras-born-free (accessed 3 February 2018).

Colla, Elliot, 2005, 'Anxious advocacy: the novel, the law, and the extrajudicial appeals in Egypt', *Public Culture* 17.3.

—, 2012, 'The people want', *The Middle East Research and Information Project* (*MERIP*), 42, https://www.merip.org/mer/mer263.

Committee to Protect Journalists, 2011, 'Egypt's reinstatement of Information Ministry is a setback', *CPJ*, 12 July, https://cpj.org/2011/07/egypts-reinstate ment-of-information-ministry-is-ma.php (accessed 25 February 2018).

Cooperson, Michael, 2000, *Classical Arabic Biography: The Heirs of the Prophets in the Age of al-Ma'mun*, Cambridge: Cambridge University Press.

Cotteret, Christophe, 2014, *Ennahdha, un histoire Tunisienne* [documentary], ARTE, http://www.tv5monde.com/programmes/fr/programme-tv-ennahdha-une-histoire-tunisienne/22801/ (accessed 28 February 2018).

Dahmani, Frida, 2012, 'Tunisie: Les salafistes, ces très inquiétants fous de Dieu', *Jeune Afrique*, 20 June, http://www.jeuneafrique.com/141108/politique/tuni sie-les-salafistes-ces-tr-s-inqui-tants-fous-de-dieu/ (accessed 16 January 2018).

—, 2014, 'Tunisie: Tension sécuritaire à trois jours des legislatives', *Jeune Afrique*, 23 October, http://www.jeuneafrique.com/Article/ARTJAWEB20141023143123/ tunisie-terrorisme-legislatives-tunisiennes-terrorisme-tunisie-tension-securtaire-a-trois-jours-des-legislatives.html (accessed 23 October 2014).

—, 2015, 'Tunisie: Le Mouvement du 18 octobre 2005, 10 ans après', *Jeune Afrique*, 28 October http://www.jeuneafrique.com/274962/politique/tunisie-le-mouve ment-du-18-octobre-2005-10-ans-apres/.

Daly, John C. K., 2017, 'Russia draws closer to Egypt', *The Arab Weekly*, 12 November, p. 17.

Daou, Marc, 2012, 'Filmé à son insu, Rached Ghannouchi tombe la masque', *France 24*, 11 October, http://www.france24.com/fr/20121011-tunisie-rached-ghannouchi-islamisme-salafistes-video-laics-polemique-ennahda (accessed 11 January 2018).

Davies, Douglas, 1995, 'The work of art in the age of digital reproduction (an evolv-ing thesis: 1991–1995)', *Leonardo* 28.5.

Davoudi, Salamander, and Ben Fenton, 2011, 'Assange signs deal with U.K. Telegraph', *The Financial Times*, 31 January, https://www.ft.com/content/0ff 84f8c-2d53-11e0-9b0f-00144feab49a?mhq5j=e6 (accessed 26 October 2017).

Derrida, Jacques, 1999, *The Gift of Death*, Chicago: University of Chicago Press.

Dickey, Christopher, 2008, 'Using comics to turn off terror', *Newsweek*, 17 April, http://www.newsweek.com/using-comics-turn-terror-86433 (accessed 1 May 2013).

Dickie, John, 2004, *Cosa Nostra*, New York: St Martin's Press.

Directinfo.com, 2012, 'Arrestation d'un Tunisien par les garde-frontières algériens pour contrebande', *Directinfo.com*, 25 October, http://directinfo.webmana gercenter.com/2012/10/25/arrestation-dun-tunisien-par-les-garde-frontieres-algeriens-pour-contrebande/ (accessed 25 October 2012).

The Economic Times, 2010, 'Qatar uses Al-Jazeera as bargaining chip: WikiLeaks', *The Economic Times*, 6 December, https://economictimes.indiatimes.com/tech/internet/qatar-uses-al-jazeera-as-bargaining-chip-wikileaks/articleshow/70 51690.cms (accessed 30 May 2018).

Elbaz, Basma, 2017, 'Stop victimizing al-Jazeera', *HuffPost*, 29 June, https://www. huffingtonpost.com/entry/stop-victimizing-al-jazeera_us_5954f466e4b0f078 efd98794 (accessed 30 May 2018).

Elmeshad, Mohamad, 2011, 'Crime on the rise, but are police to blame?', *Al Masri al-Youm*, 23 May, http://www.egyptindependent.com/crime-rise-are-police-blame/ (accessed 23 October 2017).

Emarat al-youm, 2011, '"Omar Suleiman Na'iban l-il Mubarak . . . wa Shawfiq ra'isan', *Emarat al-youm*, 30 January, http://www.emaratalyoum.com/politics/news/2011-01-30-1.348735 (accessed 24 February 2018).

Ettounsiya TV, 2014. Ali Laârayedh affirme avoir décidé de ne pas arrêter Abou Iyadh', *YouTube*, 1 June, https://www.youtube.com/watch?v=-JBzLXDIv9s (accessed 17 January 2018).

Fadl, Belal, 2011, 'Al-Baltagiya 7 najum', *Facebook*, 11 October, https://www. facebook.com/note.php?note_id=254919557888186 (accessed 14 December 2017).

—, 2014, 'The political marshal of Egypt', *Mada Masr*, 2 February, http://www. madamasr.com/en/2014/02/02/opinion/u/the-political-marshal-of-egypt/ (accessed 23 May 2017).

Fandi, Mamoun, 2008, *Hurub kalamiya*, London: Dar al-Saqi.

Farid, Samir, 2002, *Tarikh al-raqaba 'ala al-sinima fi Misr*, al-Qahira: al-Maktab al-Misri li-Tawzi' al-Matbu'at.

El Fattah, Alaa Abd, 2016, '"I was terribly wrong – writers look back at the Arab Spring five years on', *The Guardian*, 3 January, https://www.theguardian.com/books/2016/jan/23/arab-spring-five-years-on-writers-look-back (accessed 1 February 2018).

Filipov, David, 2017, 'The notorious Kremlin-linked troll farm and the Russians trying to take it down', 8 October, https://www.washingtonpost.com/world/asia_pacific/the-notorious-kremlin-linked-troll-farm-and-the-russians-trying-

to-take-it-down/2017/10/06/c8c4b160-a919-11e7-9a98-07140d2eed02_
story.html?utm_term=.0989b15174e0 (accessed 11 October 2017).

Filiu, Jean-Pierre, 2011, *The Arab Revolution: Ten Lessons from the Democratic Uprising*, New York: Oxford University Press.

—, 2015, *From Deep State to Islamic State*, New York: Oxford University Press.

Foreign Policy, 2010, 'Think again: the Internet', *Foreign Policy*, 26 April, http://foreignpolicy.com/2010/04/26/think-again-the-internet/ (accessed 8 August 2017).

Forest, Fred, 1983, *Manifeste pour une esthetique de la communication*, Brussels: Revue D'Art Contemporain, http://www.webnetmuseum.org/html/fr/expo-retr-fredforest/textes_critiques/textes_divers/3manifeste_esth_com_fr.htm#text (accessed 13 February 2018).

Foucault, Michel, 1980, *Power/Knowledge: Selected Interviews and Other Writings, 1972–1977*, New York: Pantheon Books.

France 24, 2010, 'Muwajahat bein al-shurta wa muhtajjin fi wilayat Sidi Buzid . . .', *France 24*, 19 December, http://www.france24.com/ar/20101219-tunisia-clash-poorness-young-suicide-police (accessed 17 March 2017).

Gadamer, Hans-Georg, 1975, *Truth and Method*, New York: Continuum.

al-Galad, Magdi, 2011, 'Mohamed Hassanein Heikal fi hiwar khass . . .', *Al-Masri al-Youm* (Cairo), 1 February.

—, 2014, 'Infirad', *Al-Watan*, 1 June, https://www.elwatannews.com/news/details/495659 (accessed 25 February 2018).

Gana, Nouri, 2017, 'Sons of a beach: the politics of bastardy in the cinema of Nouri Bouzid', *Cultural Politics* 13.2, 177–93.

Gartenstein-Ross, David, 2014, 'The Arab Spring and al-Qaeda's resurgence', *The Foundation for the Defense of Democracies*, http://www.defenddemocracy.org/media-hit/the-arab-spring-and-al-qaedas-resurgence/.

Gershoni, Israel, and James P. Jankowski, 1986, *Egypt, Islam, and the Arabs: The Search for Egyptian Nationhood, 1900–1930*, New York: Oxford University Press.

Gerth, J., 2005, 'Militarys information war is vast and often secretive', *The New York Times*, 11 December, http://www.nytimes.com/2005/12/11/politics/militarys-information-war-is-vast-and-often-secretive.html?_r=0 (accessed 16 October 2017).

al-Ghannouchi, Rached, 1979, 'Al-Thawra al-Iraniya thawra islamiya', *Al-Ma'rifa*, 12 February, 3.2.

—, 1993, *Al-Hurriyat al-ʿamma fi al-dawla al-Islamiyya*, Beirut: Merkaz Darasat al-Wahda al-ʿArabiyya.

—, 2005, *Al-Qadir ʿanda Ibn Taymiyya*, Saudi Arabia: al-Markaz al-Raya.

—, 2012, '*Taharrur al-shuʿub al-ʿArabiyya min al-istibdad khatwa ʿala tariq tahrir Filistin*', *Turess.com*, 17 March, https://www.turess.com/infosplus/11450 (accessed 14 June 2018).

Ghannouchi, Rached, and Olivier Ravanello, 2015, *Entretiens d'Olivier Ravanello avec Rached Ghannouchi*, Paris: Plon.

Ghazi, 2014, 'Terrorisme: L'frère de Abou Iyadh se mettre-ti-il à table', *Le Temps*, 12 October, http://www.letemps.com.tn/article/86778/terrorisme-le-fr%C3% A8re-de-abou-iyadh-se-mettra-t-il-%C3%A0-table (accessed 24 February 2018).

Ghonim, Wael, 2013, *Revolution 2.0: the Power of the People is Greater Than the People in Power: a Memoir*, Boston: Mariner Books/Houghton Mifflin Harcourt.

Glassman, James K., 2008, 'Public Diplomacy 2.0: a new approach to global engagement', speech delivered at the New America Foundation, 1 December, https://2001-2009.state.gov/r/us/2008/112605.htm (accessed 19 June 2018).

Glazzard, Andrew, 2017, *Losing the Plot: Narrative, Counter-Narrative, and Violent Extremism*, The Hague: International Centre for Counter-Terrorism, https://icct.nl/wp-content/uploads/2017/05/ICCT-Glazzard-Losing-the-Plot-May-2017.pdf.

Global Voices, 'Lina Ben Mhenni', *Global Voices*, https://globalvoices.org/author/lina-ben-mhenni/.

Goldstein, Eric, and Oumayma Ben Abdallah, 2015, 'Missing journalists: Tunisia's Arab Spring meets Libya's', *openDemocracy*, 5 May, https://www.hrw.org/news/2015/05/05/missing-journalists-tunisias-arab-spring-meets-libyas (accessed 30 January 2018).

Goodall, H. L., 2010, *Counter-narrative: How Progressive Academics can Challenge Extremists and Promote Social Justice*, Walnut Creek: Left Coast Press.

Greenberg, Nathaniel, 2010, 'War in pieces: AMIA and the triple frontier in Argentine and American discourse on terrorism', *A Contracorriente* 8.1.

—, 2011, 'A people's protest?', *The Seattle Times*, 28 January, https://www.seattle times.com/nation-world/a-peoples-protest-the-view-from-a-cairo-coffeehouse/ (accessed 24 February 2018).

—, 2012, 'The Arab constitutions: chaos and strategy', COMOPS, 1 December, http://csc.asu.edu/2012/12/01/the-arab-constitutions-2012-chaos-and-strat egy/ (accessed 24 February 2018).

—, 2013, 'Emergent public discourse and the constitutional debate in Tunisia: a critical narrative analysis', *TelosScope*, 30 December, http://www.telospress.com/emergent-public-discourse-and-the-constitutional-debate-in-tunisia-a-critical-narrative-analysis/ (accessed 27 September 2017).

—, 2014, *The Aesthetic of Revolution in the Film and Literature of Naguib Mahfouz (1952–1967)*, Lanham: Lexington Books.

—, 2016, 'Exit ISIS, stage left: fighting for laughs in Mosul and beyond', *Jadaliyya*, 16 April, http://www.jadaliyya.com/Details/33178/Exit-ISIS,-Stage-Left-Fighting-for-Laughs-in-Mosul-and-Beyond (accessed 24 February 2018).

—, 2017, 'Mythical state: aesthetics and counter-aesthetics of the Islamic State in Iraq and Syria', *The Middle East Journal of Culture and Communication* 10, 255–71, http://booksandjournals.brillonline.com/docserver/journals/18739865/10/2-3/18739865_010_02-03_s009_text.pdf?expires=1508169166&id=id&accname=id23163&checksum=2F6435FDC8346533FA49FED31FC97D47.

Greenfield, Daniel, 2012, 'Christopher Stevens feeds the crocodile', *Front Page Magazine*, 12 September, http://www.frontpagemag.com/fpm/144003/christopher-stevens-feeds-crocodile-daniel-greenfield (accessed 31 October 2017).

Guidère, Mathieu, 2014, 'The Timbuktu letters: new insights about AQIM', *Res Militaris* 4.1.

Gunthert, André, 2011, 'Photographier la révolution tunisienne' [Mp3 recording with Azyz Amami], *L'Ateliêr des icons*, http://histoirevisuelle.fr/cv/icones/1860.

Gutiérrez, Koldo, 'Arte desde la trinchera en apoyo a los maestros de Oaxaca', *Cactus*, https://www.revistacactus.com/arte-desde-la-trinchera-en-apoyo-a-los-maestros-de-oaxaca/ (accessed 26 February 2018).

Habermas, Jürgen, 1990, *The Philosophical Discourse of Modernity: Twelve Lectures*, Cambridge, MA: MIT Press.

—, 1991 [1962], *The Structural Transformation of the Public Sphere*, Cambridge, MA: MIT Press.

Haddad, Bassam, 2016, 'The debate over Syria has reached a dead end', *The Nation*, 18 October, https://www.thenation.com/article/the-debate-over-syria-has-reached-a-dead-end/ (accessed 21 July 2017).

Hadith al-thawra, 2015, 'Hal yakhsha al-Sisi 'awdat Jamal Mubarak?', *Al-Jazeera*, 5 October, http://www.aljazeera.net/programs/revolutionrhetoric/2015/5/9/%D8%A7%D9%84%D9%82%D8%B5%D9%88%D8%B1-%D8%A7%D9%84%D8%B1%D8%A6%D8%A7%D8%B3%D9%8A%D8%A9-%D8%AD%D9%83%D9%85-%D9%82%D8%B6%D8%A7%D8%A6

D9%8A-%D8%A3%D9%85-%D9%85%D9%86%D8%A7%D9%88%D8%B1%D8%A9-%D8%B3%D9%8A%D8%A7%D8%B3%D9%8A%D8%A9 (accessed 1 July 2017).

Hafez, Sabry, 2010, 'The new Egyptian novel', *New Left Review* 64, https://new-leftreview.org/II/64/sabry-hafez-the-new-egyptian-novel (accessed 1 January 2017).

Halverson, Jeffry R., and Nathaniel Greenberg, 2017, 'Ideology as narrative: the mythic discourse of al-Qaeda in the Islamic Maghrib', *Middle East Journal of Culture and Communication* 10.1.

Halverson, Jeffry R., H. L. Goodall and Steve Corman 2011, *Master Narratives of Islamic Extremism*, New York: Palgrave.

Halverson, Jeffry R., Scott W. Ruston and Angela Trethewey, 2013, 'Mediated martyrs of the Arab Spring: new media, civil religion, and narrative in Tunisia and Egypt', *Journal of Communication* 63.2, 312–32.

Hamdi, Safouane, and Yannick Bressan, 2013, 'Du théâtre à l'amphithéâtre: pour une extension du concept d'adhésion à la neuroscience éducationnelle', *Neuroéducation* 2.1.

Hamdy, Naila, 2016, 'The culture of Arab journalism', in Mohamed Zayani and Suzi Mirgani (eds), *Bullets and Bulletins: Media and Politics in the Wake of the Arab Uprisings*, Oxford: Oxford University Press.

Hamdy, N., and E. H. Gomaa, 2012, 'Framing the Egyptian uprising in Arabic language newspapers and social media', *Journal of Communication* 62.2.

Harakat al-Nahdha, 2013, 'Tada'wa ila Rachid', *Harakat al-Nahdha*, 20 March, http://www.ennahdha.tn/ (accessed 20 March 2013).

Hassan, Abdallah F., 2015, *Media, Revolution, and Politics in Egypt*, London: I. B. Tauris.

Heidegger, Martin, 1977, *Basic Writings*, New York: Harper and Row.

Herrera, Linda, 2014, *Revolution in the Age of Social Media*, London: Verso.

Hodgson, Marshall G. S., 1980, *The Order of Assassins*, New York: AMS Press.

Hope, Christopher, Robert Winnett, Holly Watt and Heidi Blake, 2011, 'WikiLeaks: Guantanamo Bay terrorist secrets revealed', *The Telegraph*, 25 April, http://www.telegraph.co.uk/news/worldnews/wikileaks/8471907/WikiLeaks-Guantanamo-Bay-terrorist-secrets-revealed.html (accessed 22 January 2018).

Howard, Phillip, and Muzammil M. Hussain, 2013, *Democracy's Fourth Wave: Digital Media and the Arab Spring*, Oxford: Oxford University Press.

Huffpost Maghreb, 2015, 'Tunisie: Le président du Conseil islamique révoqué après avoir comparé Youssef Seddik à Salman Rushdie', *Huffpost Maghreb*, 7 April,

http://www.huffpostmaghreb.com/2015/07/04/tunisie-conseil-islamique-yous
sef-seddik_n_7728078.html (accessed 8 September 2017).

Human Rights Watch, 2011, 'Tunisia: Hold police accountable for shootings',
Human Rights Watch, 29 January, https://www.hrw.org/news/2011/01/29/
tunisia-hold-police-accountable-shootings (accessed 26 February 2018).

—, 2014, 'All according to plan: the Rab'a massacre and mass killings of protesters in Egypt', *Human Rights Watch*, 12 August, https://www.hrw.org/
report/2014/08/12/all-according-plan/raba-massacre-and-mass-killings-prot
esters-egypt (accessed 6 June 2017).

Husayn, Mohamed, 2017, 'Rihlat Saʿud al-Batal', *Huffpost Arabic*, 1 December,
http://www.huffpostarabi.com/2017/11/28/story_n_18671064.html (accessed
February 2018).

Husayn, Taha, 2015, 'Adab al-thawra wa thawrat al-adab', Special reprint, *Al-Hilal*,
Cairo: Dar al-Hilal.

Ibrahim, Sarah, 2015, 'Egyptian cinema in crisis: the age of the low budget film',
Mada Masr, 20 January, https://www.madamasr.com/en/2015/01/20/feature/
culture/egyptian-cinema-in-crisis-the-age-of-the-low-budget-film/ (accessed 8
October 2017).

Iskander, Adel, 2011, 'The Baltageya: Egypt's counterrevolution', *The Huffington
Post*, www.huffingtonpost.com/adel-iskandar/the-baltageya-egypts-coun_b_
862267.html (accessed 24 October 2017).

Jackson, Nicholas, 2011, 'WikiLeaks' Assange: Facebook is an appalling spy machine',
The Atlantic, 2 May, https://www.theatlantic.com/technology/archive/2011/05/
wikileaks-assange-facebook-is-appalling-spying-machine/238225/ (accessed 24
February 2018).

Jacob, Wilson Chacko, 2007, 'Eventful transformations: al-futuwwa between history
and the everyday', *Comparative Studies in Society and History* 49.3.

Jameson, Fredric, 1971, *Marxism and Form*, Princeton: Princeton University Press.

—, 1991, *Postmodernism, or The Culture Logic of Late Capitalism*, Durham, NC:
Duke University Press.

—, 2007, *Archaeologies of the Future: The Desire Called Utopia and Other Science
Fictions*, London: Verso.

Al-Jarrah, Nouri, 1998, 'Al-ʿAzima fi sanatiha al-ʿashara', *Al-Hayat*, 5 May, http://
www.alhayat.com/article/963279/%D8%A7%D9%84%D8%A3%D8%B2%
D9%85%D8%A9-%D9%81%D9%8A-%D8%B3%D9%86%D8%AA%
D9%87%D8%A7-%D8%A7%D9%84%D8%B9%D8%A7%D8%B4%
D8%B1%D8%A9-%D8%A7%D9%84%D8%AD%D9%8A%D8%A7%

D8%A9-%D8%AA%D8%AA%D8%AC%D9%88%D9%84-%D8%B9%
D9%84%D9%89-%D9%85%D8%AF%D8%A7%D8%B1-31-%D9%8A%
D9%88%D9%85%D8%A7-%D9%81%D9%8A-%D8%A7%D9%84%
D8%AC%D8%B2%D8%A7%D8%A6%D8%B1-10-%D8%B1%D8%B3%
D8%A7%D9%85-%D8%A7%D9%84%D9%83%D8%A7%D8%B1%
D9%8A%D9%83%D8%A7%D8%AA%D9%88%D8%B1-%D8%A3%
D9%8A%D9%88%D8%A8-%D8%A7%D9%84%D9%85%D8%BA%
D8%B6%D9%88%D8%A8-%D8%B9%D9%84%D9%8A%D9%87-%
D9%85%D9%86-%D8%A7%D9%84%D8%B3%D9%84%D8%B7%D8%
A9-%D8%A7%D9%84%D8%B5%D8%AD%D8%A7%D9%81%D8%
A9-%D8%A7%D9%84%D8%AC%D8%B2%D8%A7%D8%A6%D8%
B1%D9%8A%D8%A9-%D9%85%D8%AA%D9%88%D8%A7%D8%
B7%D8%A6%D8%A9-%D9%88%D9%87%D9%8A-%D8%AA%D8%
B9%D8%B1%D9%81-%D9%85%D9%86-%D8%A7%D9%84%D8%
B0%D9%8A-%D9%8A%D9%82%D8%AA%D9%84-!-nbsp (accessed 3
May 2017).
—, 2000, *Al-Firdaws al-dami: 31 youm fi al-Jaza'ir*, Beirut: Riad el-Rayyes.
Jauss, Hans Robert, 1982, *Toward an Aesthetic of Reception*, trans. Timothy Bahti,
intro. Paul De Man, Minneapolis: University of Minnesota Press.
al-Jazairi, Abdallah, 2010, 'Forum participant warns against inciting Algerian tribes
against AQLIM' [the scum apostates endeavor to propagate sedition between
our kinfolk, the tribes and their sons of al-Qa'ida], *Al-Andalus Establishment
for Media Production*, 5 April, http://www.opensource.gov (accessed 1 October
2012).
al-Jazairi, Shaykh Abu-Muslim, 2011, 'Fatwa supports "Campaign to defend the
Islamic identity of Tunisia"', *Al-Andalus Establishment for Media Production*,
http://www.opensource.gov (accessed 18 May 2013).
Al-Jazeera, 2010, 'Tunis tahjuzu sahifat al-mawqif al-mu'arada', *Al-Jazeera*, 17 July,
http://www.aljazeera.net/news/humanrights/2010/7/17/%D8%AA%D9%
88%D9%86%D8%B3-%D8%AA%D8%AD%D8%AC%D8%B2-%D8%
B5%D8%AD%D9%8A%D9%81%D8%A9-%D8%A7%D9%84%D9%
85%D9%88%D9%82%D9%81-%D8%A7%D9%84%D9%85%D8%B9%
D8%A7%D8%B1%D8%B6%D8%A9 (accessed 16 February 2018).
—, 2011, 'Khitab Ben Ali . . .', *Al-Jazeera*, 14 January, http://www.aljazeera.net/
news/arabic/2011/1/14/%D8%AE%D8%B7%D8%A7%D8%A8-%D8%
A8%D9%86-%D8%B9%D9%84%D9%8A-%D9%8A%D8%AB%D9%
8A%D8%B1-%D8%B1%D8%AF%D9%88%D8%AF-%D8%A3%D9%

81%D8%B9%D8%A7%D9%84-%D9%85%D8%AA%D9%86%D8%A7%
D9%82%D8%B6%D8%A9 (accessed 14 June 2017).

—, 2013, 'Nass al-bayan al-quwwat al-musallaha al-Masriya', *Al-Jazeera*, 2 July,
http://www.aljazeera.net/news/reportsandinterviews/2013/7/2/%D9%86%
D8%B5-%D8%A8%D9%8A%D8%A7%D9%86-%D8%A7%D9%84%
D9%82%D9%88%D8%A7%D8%AA-%D8%A7%D9%84%D9%85%
D8%B3%D9%84%D8%AD%D8%A9-%D8%A7%D9%84%D9%85%
D8%B5%D8%B1%D9%8A%D8%A9-1-7-2013 (accessed 21 July 2017).

—, 2014, 'Harakat al-tawhid wa jihad fi gharb Afriqiya', *Al-Jazeera*, http://www.aljaz
eera.net/encyclopedia/movementsandparties/2014/2/12/%D8%AD%D8%
B1%D9%83%D8%A9-%D8%A7%D9%84%D8%AA%D9%88%D8%
AD%D9%8A%D8%AF-%D9%88%D8%A7%D9%84%D8%AC%D9%
87%D8%A7%D8%AF-%D9%81%D9%8A-%D8%BA%D8%B1%D8%
A8-%D8%A3%D9%81%D8%B1%D9%8A%D9%82%D9%8A%D8%A7
(accessed 9 January 2018).

Jilasi, Fatima, 2014, 'Takshifu asrar wa ahdaf ta'sis Ansar al-Shari'a', *Al-Sabah*, 6
March, http://www.assabah.com.tn/article/80924/%D8%A7%D9%84%D8%
B5%D8%A8%D8%A7%D8%AD-%D8%AA%D9%83%D8%B4%D9%
81-%D8%A3%D8%B3%D8%B1%D8%A7%D8%B1-%D9%88%D8%
A7%D9%87%D8%AF%D8%A7%D9%81-%D8%AA%D8%A3%D8%
B3%D9%8A%D8%B3-%D8%A7%D9%86%D8%B5%D8%A7%D8%
B1-%D8%A7%D9%84%D8%B4%D8%B1%D9%8A%D8%B9%D8%A9
(accessed 6 March 2014).

al-Jourshi, Salah al-Din, 2011, 'Al-Mushahid al-Islami fi Tunis: qawiy wa muw-
waqif', in *Min Qabdat Bin Ali Ila thawrat al-yasamin: al-islam al-Siyasi fi Tunus*,
Abu Dhabi: Al-Mesbar Studies & Research Center.

Al-Jumhuriya, 2011, 'Makasib al-thawra al-sha'biya', *Al-Jumhuriya* (Cairo), 4
February, p. 1.

Kahlaoui, Tarek, 2013, 'The powers of social media', in Nouri Gana (ed.), *The
Making of the Tunisian Revolution*, Edinburgh: Edinburgh University Press.

Kallander, Amy Aisen, 2013, 'From TUNeZINE to Nhar 3la 3mmar: a reconsideration of the role of bloggers in Tunisia's revolution', *Arab Media & Society*, 27
January, http://www.arabmediasociety.com/?article=818 (accessed 17 February
2018).

Kapitalis, 2015, 'Un témoin libyen confirme l'assassinat de Sofiane et Nadhir',
Kapitalis, 8 January, http://kapitalis.com/tunisie/2017/01/08/un-temoin-
libyen-confirme-lassassinat-de-sofiane-et-nadhir/ (accessed 25 July 2017).

Al-Karama, 2007, 'Tunisia: risk of violation of Sayfallah Ben Hassine's right to life', *Al-Karama for Human Rights*, 17 July, https://www.alkarama.org/en/articles/tunisia-risk-violation-sayfallah-ben-hassines-right-life (accessed 19 January 2018).

Karawya, Fayrouz, 2010, 'Studio Misr to be auctioned off today', *Al-Masri al-Youm*, 18 July, http://www.egyptindependent.com/news/studio-misr-be-auctioned-today (accessed 10 May 2017).

Kelmetna, 2011, 'Mahmoud ʿAthman . . . al-muharraḍ ʿala al-thawra', *Kelmetna*, 12 March, http://www.masress.com/kelmetna/10505 (accessed 28 February 2018).

Kendall, Elisabeth, 2010, *Literature, Journalism and the Avant-garde: Intersection in Egypt*, London: Routledge.

al-Khatib, Ahmed, 2014, 'Kol ma yajib taʿrifuhu ʿan ʿhurub al-jil al-rabiʿ', *Sasapost*, 26 August, http://www.sasapost.com/4th-generation-warfare/ (accessed 21 July 2017).

Khatib, Lina, W. Dutton and M. Thelwall, 2012, 'Public Diplomacy 2.0: a case study of the US Digital Outreach Team', *Middle East Journal* 66.3, 453–72.

Khayat, M, 2012, 'The rise of the Salafi-Jihadi movement in Tunisia – the case of Ansar Al-Shariʿa in Tunisia', *The Middle East Media Research Institute*, 20 March, http://www.memrijttm.org/the-rise-of-the-salafi-jihadi-movement-in-tunisia--the-case-of-ansar-al-sharia-in-tunisia.html (accessed 3 January 2018).

Kirkpatrick, David D., 2014, 'Prolonged fight feared in Egypt after bombings', *The New York Times*, 24 January, https://www.nytimes.com/2014/01/25/world/middleeast/fatal-bomb-attacks-in-egypt.html (accessed 24 February 2018).

—, 2017, 'In snub to U.S., Russia and Egypt move toward deal on air base', *The New York Times*, 30 November, https://www.nytimes.com/2017/11/30/world/middleeast/russia-egypt-air-bases.html (accessed 25 February, 2018).

Kishtainy, Khalid, 1985, *Arab Political Humor*, London: Quartet Books.

Kraidy, Marwan M., 2016, *The Naked Blogger of Cairo: Creative Insurgency in the Arab World*, Cambridge, MA: Harvard University Press.

Kranish, Michael, Tom Hamburger and Carol D. Leonnig, 2017, 'Michael Flynn's role in Mideast nuclear project could compound legal issues', *The Washington Post*, 27 November, https://www.washingtonpost.com/politics/michael-flynns-role-in-middle-eastern-nuclear-project-could-compound-legal-issues/2017/11/26/51ce7ec8-ce18-11e7-81bc-c55a220c8cbe_story.html?utm_term=.b53af7eb779a (accessed 25 February 2018).

Krichen, Aziz, 2016, *La Promesse au printemps*, Paris: Script.

Laclau, Ernesto, 2005, *On Populist Reason*, New York: Verso.

Lacroix, Stéphane, 2012, 'Le salafisme, "c'est le dogme dans toute sa pureté"', *Le Monde*, 27 September, http://www.lemonde.fr/culture/article/2012/09/27/le-dogme-dans-toute-sa-purete_1766968_3246.html#yYzm5pyIwbviHdAO.99 (accessed 1 October 2013).

Landler, Mark, and Andrew W. Lehren, 2011, 'Cables show delicate U.S. dealings with Egypt's leaders', *The New York Times*, 27 January, http://www.nytimes.com/2011/01/28/world/middleeast/28diplo.html?_r=0 (accessed 11 October 2017).

László, János, 2008, *The Science of Stories: An Introduction to Narrative Psychology*, New York: Routledge.

Lecomte, Roman, 2009, 'Internet et la reconfiguration de l'espace public tunisien: le rôle de la diaspora', *Nawaat*, 20 December, https://nawaat.org/portail/2009/12/20/internet-et-la-reconfiguration-de-lespace-public-tunisien-le-role-de-la-diaspora/ (accessed 24 July 2017).

Ledwith, Sara, 2009, 'Iran's Neda shows citizen journalism unleashed', *Reuters*, 23 June, https://www.reuters.com/article/us-internet-iran-analysis-sb/irans-neda-shows-citizen-journalism-unleashed-idUKTRE55M39820090623 (accessed 18 June 2018).

Levinas, Gabriel, 1998, *La Ley Bajo Los Escombros: AMIA: Lo Que No Se Hizo*, Buenos Aires: Editorial Sudamericana.

Lim, Merlyna, 2013, 'Framing Bouazizi: "White lies", hybrid network, and collective/connective action in the 2010–11 Tunisian uprising', *Journalism* 14.7.

Lincoln, Bruce, 2012, *Gods and Demons, Priests and Scholars: Critical Explorations in the History of Religions*, Chicago: University of Chicago Press.

Lister, Tim, 2011, 'U.S. cables: Mubarak still a vital ally', *CNN*, 28 January, http://www.cnn.com/2011/WORLD/africa/01/28/egypt.wikileaks.cables/ (accessed 11 October 2017).

Litvin, Margaret, 2013, 'From Tahrir to "Tahrir": some theatrical impulses toward the Egyptian uprising', *Theater Research International* 38.2.

Mackey, Robert, and Liam Stack, 2012, 'Obscure film mocking Muslim prophet sparks anti-U.S. protests in Egypt and Libya', *The New York Times*, 11 September, https://thelede.blogs.nytimes.com/2012/09/11/obscure-film-mocking-muslim-prophet-sparks-anti-u-s-protests-in-egypt-and-libya/?_r=0 (accessed 27 October 2017).

MacKinnon, Rebecca, 2012 *Consent of the Networked: The Worldwide Struggle For Internet Freedom*, New York: Basic Books.

Mada Masr, 2014, 'Renowned novelist Aswany quits column, citing censorship',

Mada Masr, 24 June, http://www.madamasr.com/en/2014/06/24/news/u/ renowned-novelist-aswany-quits-writing-column-citing-censorship/ (accessed 23 May 2017).

Madrigal, Alexis, 2011, 'Egyptian activists action plan: translated', *The Atlantic*, 27 January, https://www.theatlantic.com/international/archive/2011/01/egyptian-activists-action-plan-translated/70388/ (accessed 15 October 2017).

Mag 14, 2013, 'La Tunisie cède 15% de Tunisiana à Qatar Telecom', *Mag 14*, 1 January, http://www.mag14.com/capital/62-entreprises/1366-la-tunisie-cede-15-de-tunisiana-a-qatar-telecom.html (accessed 27 February 2018).

Mahfuz, Najib, 2006, *Awlad haratina*, Al-Qahira: Dar al-Shorouk.

Makhlouf, Zuheir, 2011, 'Muhawala intihar al-sha'b al-majaz Mohamed al-Bou'azizi asil Sidi Bouzid', *Assabilonline*, 17 December, https://www.facebook.com/168736489831044/videos/119774181422660/ (accessed 29 June 2018).

Malka, Haim, 2015. 'Tunisia: confronting extremism', in J. B. Alterman (ed.), *Religious Radicalism After the Arab Uprisings*, Lanham: Rowman and Littlefield, p. 115.

Mansour, Ahmed, 2012, 'Lena Ben Mhenni. Shahada 'ala al-thawra al-Tunisiya. Shahid 'ala 'asar', *Al-Jazeera*, 25 July, http://www.aljazeera.net/programs/centurywitness/2012/7/25/%D9%84%D9%8A%D9%86%D8%A7-%D8%A8%D9%86-%D9%85%D9%87%D9%86%D8%A7-%D8%B4%D8%A7%D9%87%D8%AF%D8%A9-%D8%B9%D9%84%D9%89-%D8%A7%D9%84%D8%AB%D9%88%D8%B1%D8%A9-%D8%A7%D9%84%D8%AA%D9%88%D9%86%D8%B3%D9%8A%D8%A9-%D8%AC2 (accessed 19 February 2018).

Marketwired, 2010, 'Narus wins award for best real-time dynamic network analysis and forensics solution', *Marketwired*, 9 November, http://www.marketwired.com/press-release/narus-wins-award-for-best-real-time-dynamic-network-analysis-and-forensics-solution-nyse-ba-1349435.htm (accessed 7 February 2018).

Markham, James M., 1980, 'Tunisian regime, after major setbacks, regains vitality', *The New York Times*, 21 May.

Marzouki, Nadia, 2013, 'From resistance to governance: the category of civility in the political theory of Tunisian Islamists', in Nouri Gana (ed.), *The Making of the Tunisian Revolution: Contexts, Architects, Prospects*, Edinburgh: Edinburgh University Press.

Al-Masri al-youm, 2011, 'Harb al-shawari' fi al-Qahira', *Al-Masri al-youm* (Cairo), 29 January.

—, 2011, 'Al-Nida' al-akhir . . .', *Al-Masri al-youm* (Cairo), 29 January.

McGrath, Dermott, 2002, 'Tunisian net dissident jailed', *Wired*, 13 June, https://www.wired.com/2002/06/tunisian-net-dissident-jailed/ (accessed 24 July 2017).

McTighe, Kristen, 2011, 'A blogger at Arab Spring's genesis', *The New York Times*, 12 October, http://www.nytimes.com/2011/10/13/world/africa/a-blogger-at-arab-springs-genesis.html.

Mekay, Emad, 2011, 'TV stations multiply as Egyptian censorship falls', *The New York Times*, 30 July, http://www.nytimes.com/2011/07/14/world/middleeast/14iht-M14B-EGYPT-MEDIA.html (accessed 7 June 2017).

Mellor, Noha, 2007, *Modern Arab Journalism*, Edinburgh: Edinburgh University Press.

—, 2016, *The Egyptian Dream: Egyptian National Identity and Uprisings*, Edinburgh: Edinburgh University Press.

—, 2018, *Voice of the Muslim Brotherhood: Da'wa, Discourse, and Political Communication*, London: Routledge.

Merone, Fabio, 2013, 'Salafism in Tunisia: an interview with a member of Ansar al-Sharia', *Jadaliyya*, 11 April, http://www.jadaliyya.com/pages/index/11166/salafism-in-tunisia_an-interview-with-a-member-of- (accessed 24 February 2018).

Michaelson, Ruth, 2017, 'Egypt blocks access to news websites including Al-Jazeera and Mada Masr', *The Guardian*, 25 May, https://www.theguardian.com/world/2017/may/25/egypt-blocks-access-news-websites-al-jazeera-mada-masr-press-freedom (accessed 20 July 2017).

Minbar al-Jazeera, 2010, 'Hajb al-mawaqi' al-iliktruniya', *Al-Jazeera*, 24 May, http://www.aljazeera.net/programs/al-jazeera-platform/2010/5/24/%D8%AD%D8%AC%D8%A8-%D8%A7%D9%84%D9%85%D9%88%D8%A7%D9%82%D8%B9-%D8%A7%D9%84%D8%A5%D9%84%D9%83%D8%AA%D8%B1%D9%88%D9%86%D9%8A%D8%A9 (accessed 17 February 2018).

Le Monde, 2011, 'L'Egypt: autre "ennemi d'Internet"', *Le Monde*, 26 January, http://www.lemonde.fr/technologies/article/2011/01/26/l-egypte-autre-ennemi-d-internet_1470630_651865.html (accessed 13 July 2017).

Moore, Henry Clement, 1965, *Tunisia Since Independence*, Berkeley: University of California Press.

—, 1973, 'Tunisia after Bourguiba', in I. William Zartman (ed.), *Man, State, and Society in the Contemporary Maghreb*, New York: Praeger.

Morsy, Ahmed, 2013, 'Rough justice', *Al-Ahram Weekly*, 21 March, http://weekly. ahram.org.eg/News/1946.aspx (accessed 23 October 2017).

Mosaique FM, 2014, 'Abou Iyadh se cachait à proximité du ministère de l'Intérieur', *Mosaique FM*, 3 March, http://archivev2.mosaiquefm.net/fr/index/a/ ActuDetail/Element/35191-abou-iyadh-se-cachait-a-proximite-du-ministere-de-l-interieur (accessed 17 January 2018).

—, 2014, 'Bin Jidu: Abu 'Ayadh fi Derna', *Mosaique FM*, 29 November.

—, 2017, 'Sofiane et Nédhir vivants: Bghouri émet des doutes', *Mosaique FM*, 16 July, http://www.mosaiquefm.net/fr/actualite-national-tunisie/172373/sofiane-et-nedhir-vivants-bghouri-emet-des-doutes (accessed 30 January 2018).

Mouloud, Abdullah, 2017, 'Aymin 'Aam "harakat tahrir Azwad"', *Al-Quds al-'arabi*, 13 May, http://www.alquds.co.uk/?p=718667 (accessed 9 January 2018).

Mourad, Mahmoud, 2015, 'Egyptian poet goes on trial accused of contempt of Islam', *Reuters*, 28 January, http://www.reuters.com/article/us-egypt-courts-poet-idUSKBN0L121M20150123 (accessed 23 May 2017).

Munoz, Arturo, 2012, *U. S. Military Information Operations in Afghanistan: Effectiveness of Psychological Operations 2001–2010*, Santa Monica: RAND.

Murphy, Dennis M., and James F. White, 2007, *Propaganda: Can a Word Decide a War?* The US Army War College, http://www.dtice.mil/docs/citations/ ADA486008 (accessed 19 June 2018).

Mustafa, Hani, 2012, 'Shantytown Dogs', *Al-Ahram Weekly*, 15 July, http://weekly. ahram.org.eg/Archive/2012/1107/cu3.htm (accessed 23 October 2017).

National Commission on Terrorist Attacks upon the United States, 2004, *The 9/11 Commission Report: Final Report of the National Commission on Terrorist Attacks Upon the United States*, New York: Norton.

Nawaat, 2010, 'TuniLeaks, les documents dévoilés par WikiLeaks concernant la Tunisie: Quelques réactions à chaud', *Nawaat*, 28 November, http://nawaat. org/portail/2010/11/28/tunileaks-les-documents-devoiles-par-wikileaks-concernant-la-tunisie-quelques-reactions-a-chaud/ (accessed 17 February 2018).

—, 2011, 'A la mémoire de Mohamed Bouazizi', *Nawaat*, 6 January, https://nawaat. org/portail/2011/01/06/a-la-memoire-de-mohamed-bouazizi/ (accessed 16 March 2017).

The Observers, 2013, 'Tunisian hackers decrypt dictator's old Internet censorship machines', *France 24*, 24 June, http://observers.france24.com/en/20130624-tunisia-internet-censorship-hackers-servers (accessed 24 July 2017).

ONTV, 2012, ''Alaqat Amrika bi-il-ikhwan al-muslimin', *ONTV*, 31 December, https://www.youtube.com/watch?v=QCKwsa8KuEk (accessed 2 November 2017).

Osman, Tarek, 2011, *Egypt on the Brink*, New Haven: Yale University Press.

Osnos, David, 2017, 'The new Cold War', *The New Yorker*, 6 March, https://www.newyorker.com/magazine/2017/03/06/trump-putin-and-the-new-cold-war (accessed 24 February 2018).

Pargeter, Alison, 2012, 'Radicalisation in Tunisia', in George Joffe (ed.), *Islamist Radicalisation in North Africa: Politics and Process*, London: Routledge.

Parikka, Jussi, 2007, *Digital Contagions*, New York: Peter Lang.

Paris, Gilles, 2012, 'Le salafisme, c'est "le dogme dans toute sa pureté"', *Le Monde*, 27 September, http://www.lemonde.fr/culture/article/2012/09/27/le-dogme-dans-toute-sa-purete_1766968_3246.html#yYzm5pyIwbviHdAO.99 (accessed 1 February 2018).

Perkins, Kenneth, 2013, 'Playing the Islamist card: the use and abuse of religion in Tunisian politics', in Nouri Gana (ed.), *The Making of the Tunisian Revolution*, Edinburgh: Edinburgh University Press.

Plato, 1992, *The Republic*, trans. G. M. A. Grube and C. D. C. Reeve, Indianapolis: Hackett Publishing Co.

Policy Coordinating Committee on Public Diplomacy and Strategic Communication (US), 2007, *U.S. National Strategy for Public Diplomacy and Strategic Communication*, pp. 4–5, http://www.au.af.mil/au/awc/awcgate/state/natstrat_strat_comm.pdf.

POMED, 2011, 'The Weekly Wire', *POMED*, 25 April, http://www.pomed.org/wp-content/uploads/2011/05/Weekly-Wire-April-25.pdf (accessed 8 August 2017).

Pomerantsev, Peter, 2014a, *Nothing is True and Everything is Possible*, New York: Public Affairs.

—, 2014b, 'Russia and the menace of unreality', *The Atlantic*, 9 September, https://www.theatlantic.com/international/archive/2014/09/russia-putin-revolutionizing-information-warfare/379880/ (accessed 24 February 2018).

Prince, Erik, 2016, 'Obama and Clinton are complicit in creating ISIS', *Breitbart.com*, 16 June, http://www.breitbart.com/national-security/2016/06/16/erik-prince-obama-clinton-complicit-creating-isis/ (accessed 25 January 2018).

Qandil, Abd al-Halim, 2008, *Al-Ayyam al-Akhira*, al-Qahira: Dar al-Thaqafa al-Jadida.

—, 2014, *Hukm al-Jiniralat*, al-Qahira: Maktabat Jazirat al-Ward.

Qandil, Hamdi, 2014, *Ishtu marratayn*, Cairo: Dar al-Shorouk.

Rama, Ángel, 2009, *La ciudad letrada*, México: Fineo.

Ramadan, Ayman, 2017, 'Wa'il al-Ibrashi', *Youm 7*, 20 October, http://www.youm7.com/story/2014/10/20/%D9%88%D8%A7%D8%A6%D9%84-%D8%A7%D9%84%D8%A5%D8%A8%D8%B1%D8%A7%D8%B4%D9%89-%D9%88%D9%82%D9%81-%D8%A7%D9%84%D8%A8%D8%AB-%D8%B5%D9%86-%D8%A7%D9%84%D8%B9%D8%A7%D8%B4%D8%B1%D8%A5-%D9%85%D8%B3%D8%A7%D8%A1-%D9%84%D8%A3%D8%B1%D8%A8%D8%A7%D8%A8-%D8%B3%D9%8A%D8%A7%D8%B3%D9%8A%D8%A9-%D9%88%D9%84%D9%8A%D8%B3%D8%AA/1913602 (accessed 23 May 2017).

Rancière, Jacques, 2004, *The Politics of Aesthetics: the Distribution of the Sensible*, London: Continuum.

—, 2007, *On the Shores of Politics*, London: Verso.

Raphael, Gael, 2011, 'Al-Shaabi's "The Will to Life"', *Jadaliyya*, 2 May, http://www.jadaliyya.com/Details/23935/Al-Shabbi%60s-The-Will-to-Life (accessed 11 January 2018).

Reporters without Borders, 2002, 'Imprisoned cyber-dissident risks five-year jail sentence', *Reporters without Borders*, 12 June, https://rsf.org/en/news/imprisoned-cyber-dissident-risks-five-year-jail-sentence (accessed 24 July 2017).

Reuters, 2010, 'Witnesses report rioting in Tunisian town', *Reuters*, 19 December, http://www.reuters.com/article/ozatp-tunisia-riot-idAFJOE6BI06U20101219 (accessed 17 March 2017).

RFI, 2015, 'Le groupe EI en Libye dit avoir tué deux journalistes tunisiens', *RFI*, 8 January, http://www.rfi.fr/afrique/20150108-branche-libyenne-ei-annonce-assassinat-chourabi-guetari/ (accessed 3 January 2016).

Ricoeur, Paul, 2005, *The Course of Recognition*, Cambridge, MA: Harvard University Press.

Roberts, Ellery Biddle, and Afef Abrougui, 2017, 'Tunisian media activist interrogated over sources of leaked documents', *Global Voices*, 8 May, https://advox.globalvoices.org/2017/05/08/tunisian-media-activist-interrogated-over-sources-of-leaked-documents/ (accessed 14 February 2018).

Roger, Benjamin, 2013, 'Tunisie: sur la trace des jihadistes du mont Chaambi', *Jeune Afrique*, 5 July, http://www.jeuneafrique.com/Article/ARTJAWEB20130507161818/alg-rie-tunisie-terrorisme-al-qaeda-terrorisme-tunisie-sur-la-trace-des-jihadistes-du-mont-chaambi.html (accessed 5 July 2013).

Ross, L., G. Bierbauer and S. Hoffman, 1976, 'The role of attribution processes in conformity and dissent: revisiting the Asch situation', *American Psychologist* 31, 148–57.

Ross, Tim, Mathew Moore and Steven Swinford, 2011, 'Egypt protests: America's secret backing for rebel leaders behind uprising', *The Daily Telegraph*, 28 January, http://www.telegraph.co.uk/news/worldnews/africaandindianocean/egypt/8289686/Egypt-protests-Americas-secret-backing-for-rebel-leaders-behind-uprising.html (accessed 11 October 2017).

Rostain, Sophie, 2005, 'L'Islam à Coran ouvert', *Libération*, 22 April, http://www.liberation.fr/medias/2005/04/22/l-islam-a-coran-ouvert_517419 (accessed 8 September 2017).

Rouissi, Slimane, 2010, 'Public suicide sparks angry riots in central Tunisia', *The Observers*, 21 December, http://observers.france24.com/en/20101221-youth-public-suicide-attempt-sparks-angry-riots-sidi-bouzid-tunisia-poverty-bouazizi-immolation (accessed 6 August 2017).

—, 2012, 'Connaissez-vous Slimane Rouissi: l'homme qui a lance la revolution Tunisienne', *France 24*, 17 December, http://observers.france24.com/fr/20111216-connaissez-vous-slimane-rouissi-l%E2%80%99homme-lance-revolution-tunisienne-sidi-bouzid-bouazizi-14-janvier (accessed 6 August 2017).

RT, 2013, 'Isra'il wa Suriya wa Iran: milaffat Kerry al-hayawiya l-il-Ikhwan Misr', *RT*, 4 March, https://arabic.rt.com/news/609314-%D8%A5%D8%B3%D8%B1%D8%A7%D8%A6%D9%8A%D9%84_%D9%88%D8%B3%D9%88%D8%B1%D9%8A%D8%A7_%D9%88%D8%A5%D9%8A%D8%B1%D8%A7%D9%86_%D9%85%D9%84%D9%81%D8%A7%D8%AA_%D9%83%D9%8A%D8%B1%D9%8A_%D8%A7%D9%84%D8%AD%D9%8A%D9%88%D9%8A%D8%A9_%D9%84%D8%A5%D8%AE%D9%88%D8%A7%D9%86_%D9%85%D8%B5%D8%B1/ (accessed 15 November 2017).

—, 2015, 'Al-Jaysh al-Suri yath'ar l-il-Kisabsa', *RT*, 10 March, https://arabic.rt.com/news/795736-%D8%AC%D9%8A%D8%B4-%D8%B3%D9%88%D8%B1%D9%8A-%D8%A3%D8%A8%D9%88-%D8%A8%D9%84%D8%A7%D9%84-%D8%A7%D9%84%D8%AA%D9%88%D9%86%D8%B3%D9%8A-%D8%BA%D8%A7%D8%B1%D8%A9-%D8%AC%D9%88%D9%8A%D8%A9/ (accessed 24 February 2018).

Ryan, Yasmine, 2011, 'Tunisia's bitter cyberwar', *Al-Jazeera*, 6 January, http://www.aljazeera.com/indepth/features/2011/01/20111614145839362.html (accessed 6 January 2012).

Sabea, Hanan, 2014, '"I dreamed of being a people": Egypt's revolution, the people and critical imagination', in P. Werbner, M. Webb and K. Spellman-Poots (eds), *The Political Aesthetics of Global Protest: the Arab Spring and Beyond*, Edinburgh: Edinburgh University Press, pp. 67–92.

Sageman, Marc, 2004, *Understanding Terror Networks*, Philadelphia: University of Philidelphia Press.

Said, Edward, 2013, 'Introduction' to the Fiftieth Anniversary Edition of Erich Auerbach's *Mimesis: The Representation of Reality in Western Literature*, Princeton: Princeton University Press.

Said, Mekkawi, and 'Amr Kafrawi, 2014, *Kurrasat al-tahrir: hikayat wa-amkina*, Cairo: Dar al-Misri al-Libnaniya.

de Saint Perier, Laurent, and Leïla Slimani, 2012, 'Le Qatar investi au Maghreb', *Jeune Afrique*, 21 February, http://www.jeuneafrique.com/142844/archives-thematique/le-qatar-investit-au-maghreb/ (accessed 27 February 2018).

Salazar, Philippe-Joseph, 2017, *Words are Weapons: Inside ISIS's Rhetoric of Terror*, New Haven: Yale University Press.

Sandels, Alexandra, 2003, 'A dark year for press freedom in Egypt', *Menassat*, 9 January, http://www.menassat.com/?q=en/news-articles/2641-dark-year-press-freedom-egypt (accessed 2 January 2018).

Saraya, Osama, 2011, 'Tanzim irhabi min 19 intihariyan li-tafjir dur al-'abadah', *Al-Ahram* (Cairo), 25 January.

Schmidle, R., *et al.* 2015, *White Paper on Social and Cognitive Neuroscience Underpinnings of isil Behavior and Implications for Strategic Communication, Messaging, and Influence*, May, https://info.publicintelligence.net/SMA-ISIL-MessagingInfluence.pdf (accessed 15 October 2017).

Schwedler, Jillian, 2016. 'Taking time seriously: temporality and the Arab uprisings', *Project on Middle East Political Science*, 3–4 May, https://pomeps.org/2016/06/10/taking-time-seriously-temporality-and-the-arab-uprisings/#_ftn7 (accessed 1 February 2017).

Seddik, Youssef, 2005, *Qui sont les barbares? Itinéraire d'un penseur d'islam*, Paris: Editions l'Aube.

—, 2011–13, multiple articles in the column 'La chronique de Youssef Seddik', *Le Temps*, http://www.letemps.com.tn (accessed 1 July 2013).

—, 2012, 'La chronique de Youssef Seddik: Encore une lettre à un(e) inconnu(e) intelligent(e) parmi nos gouvernants provisoires', *Le Temps*, 25 August, http://www.letemps.com.tn/article-69172.html (accessed 1 July 2013).

—, 2013, *Nous n'avons jamais lu le Coran*, Paris: l'Aube.

—, 2014, *Tunisie: La revolution inachevée. Entretiens avec Gilles Vanderpooten*, Paris: Editions de l'Aube.

—, 2016a, *Le grand malentendu: l'Occident face au Coran*, La Tour-d'Aigues: Editions de l'Aube.

—, 2016b, 'Al-Mufakkir al-Tunisi Youssef Seddik', *Al-Araby TV*, 7 May, https://www.youtube.com/watch?v=geHa5fRUSWA (accessed 8 September 2017).

Sen, Amartya Kumar, 2006, *Identity and Violence: the Illusion of Destiny*, New York: Norton.

al-Shabi, Sabah, 2014, 'Abu 'Ayadh', *Al-Sabah*, 18 October, http://www.assabah-news.tn/article/92944/%D8%B4%D9%82%D9%8A%D9%82%D9%87-%D9%8A%D9%81%D8%AC%D9%91%D8%B1-%D9%85%D9%81D8%A7%D8%AC%D8%A3%D8%A9-%D9%85%D9%86-%D8%A7D9%84%D9%88%D8%B2%D9%86-%D8%A7%D9%84%D8%ABD9%82%D9%8A%D9%84-%D8%A3%D8%A8%D9%88-%D8%B9D9%8A%D8%A7%D8%B6-%D9%8A%D8%B1%D8%B3%D9%84-%D8%A7%D9%84%D8%A3%D9%85%D9%88%D8%A7%D9%84-%D9%85D9%86-%D9%84%D9%8A%D8%A8%D9%8A%D8%A7-%D9%81%D9%8A-%D8%B9%D9%84%D8%A8-%D8%A7%D9%84%D8%B4%D9%88%D9%83%D9%88%D9%84%D8%A7%D8%B7%D8%A9 (accessed 26 January 2015).

Sharawai, Fatima, 2015, 'Shabakat al-birnamij al-'amma', *Al-Ahram*, 16 September, http://gate.ahram.org.eg/News/751664.aspx (accessed 25 February 2018).

Sharif, Nihad, 1972, *Qahir al-zaman*, Cairo: Dar al-Hilal.

Sharuni, Yusuf, 2000, *Al-Khayal al-'ilmi fi al-adab al-'arabi al-mu'asir: hatta nihayat al-qarn al-'ishrin*, Cairo: al-Haya al-Misriyah al-'Ammah lil-Kitab.

—, 2006, 'Adab al-khayal al-'ilmi al-'arabi', *Al-Ghrad*, 26 February, http://www.alghad.com/articles/772434-%D8%A3%D8%AF%D8%A8-%D8%A7%D9%84%D8%AE%D9%8A%D8%A7%D9%84-%D8%A7%D9%84%D8%B9%D9%84%D9%85%D9%8A-%D8%A7%D9%84%D8%B9%D8%B1%D8%A8%D9%8A-%D9%8A%D8%AD%D8%AA%D8%A7%D8%AC-%D8%AA%D9%82%D8%AF%D9%85%D8%A7-%D8%AA%D9%82%D9%86%D9%8A%D8%A7-%D9%8A%D9%86%D8%AA%D8%AC-%D9%84%D8%BA%D8%A9-%D9%85%D8%BA%D8%A7%D9%8A%D8%B1%D8%A9. (accessed 26 January 2015).

El-Shobaki, Amr, 2014, 'Egyptian state at risk without reform in 2015', *Al-Ahram Online*, 29 December, http://english.ahram.org.eg/NewsContent/1/64/119048/Egypt/Politics-/INTERVIEW-Egyptian-state-at-risk-without-reform-in.aspx (accessed 26 February 2018).

—, 2016, 'Madha tabaqqa min thawrat yanayir?!', *Majallat al-dimuqratiya al-Ahram* 61, 27.

Al-Shorouk [Al-Shoruq], 2011, 'Youm al-ghadab', *Al-Shorouk* (Cairo), 25 January.

—, 2011, 'Burkan al-ghadab yejtah shawari' al-Qahira', *Al-Shorouk* (Cairo), 26 January.

—, 2011, ''Unf 'ashwa'i wa qaswa amniya mufirta . . .', *Al-Shorouk* (Cairo), 27 January.

—, 2011, 'Heikal yatakalam 'ama yajri fi Misr al-an', *Al-Shorouk* (Cairo), 29 January.

—, 2011, 'Masirat l-mi'at al-alalaf min al-mutazahirin tawasul al-hitaf 'al-sha'b yurid isqat al-nizam', *Al-Shorouk* (Cairo), 30 January, p. 1.

—, 2011, 'Al-Sha'b yataqaddam wa Mubarak yabda' al-taraju'', *Al-Shorouk* (Cairo), 30 January, p. 1.

—, 2011, 'Tard 'Izz min al-watani wa anba' 'an hurub 'Alaa' wa Gamal Mubarak ila London', *Al-Shorouk* (Cairo) 30 January, p. 1.

—, 2011, 'Muqtarahat Omar Suleiman l-ihtwa' intifadhat al-ghadib', *Al-Shorouk* (Cairo), 31 January. p. 1.

—, 2011, 'Al-baltagiya yu'alinur al-harban 'ala al-taghyir', *Al-Shorouk* (Cairo), 3 February, p. 1.

—, 2011, 'Mohamed Hassenein Heikal yatakallam l-il-Shorouk', *Al-Shorouk* (Cairo), 3 February, p. 1.

—, 2011, '30 Musallahan . . .', *Al-Shorouk* (Cairo), 3 February, p. 1.

—, 2011, 'Al-dakhiliya taqul: in jihazan sirriyan laysa tabi'an laha wara' al-fauda', *Al-Shorouk* (Cairo), 5 February, p. 1.

Shawikh, Abdallah, 2017, 'Ahmed Khaled', *Emarat al-youm*, 24 March, http://www.emaratalyoum.com/life/culture/2017-03-25-1.981006 (accessed 27 February 2018).

El Sheikh, Mayy, and David Kirkpatrick, 2013, 'Rise in sexual assaults in Egypt sets off clash over blame', *The New York Times*, 25 March, http://www.nytimes.com/pages/world/index.html (accessed 25 March 2013).

Shukrallah, Hani, 2012, 'Covering the Arab Spring: myths, lies, and truths', Issam Ferres Institute for Public Policy and International Affairs, *The American University in Beruit*, 21 May, https://www.youtube.com/watch?v=Wu3nwHhAI2U (accessed 1 June 2016).

Shukrallah, Salma, 2013, 'Once election allies, Egypt's Fairmont opposition turn against Morsi', *Ahram Online*, 27 June, http://english.ahram.org.eg/

NewsContent/1/152/74485/Egypt/Morsi,-one-year-on/-Once-election-allies,-Egypts-Fairmont-opposition-.aspx (accessed 24 May 2017).

Shukri, Ghali, 1981, *Egypt, Portrait of a President, 1971–1981: the Counter-revolution in Egypt, Sadat's Road to Jerusalem*, London: Zed Press.

Sisco, Jim, 2015, 'Should the Gulf states be wary of the revolution tsunami coming their way', *The Express Tribune*, 23 August, https://blogs.tribune.com.pk/story/29134/should-the-gulf-states-be-wary-of-the-revolution-tsunami-coming-their-way/ (accessed 20 February 2018).

al-Soltani, Abd al-Latif, 1974, *Al-Mazdakiya hiya ʿasal al-Ishtirakiya*, Casablanca: Aljazaʾir, Jamiʿa al-haquq mahfuza l-il-muʾlif.

de Soto, Hernando, 2011, 'The real Mohamed Bouazizi', *Foreign Policy*, 16 December, http://foreignpolicy.com/2011/12/16/the-real-mohamed-bouazizi/ (accessed 27 September 2017).

Souheif, Ahdaf, 2011, 'Forward', in Nadia Idle and Alex Nunes (eds), *Tweets from Tahrir: Egypt's Revolution as it Unfolded from the People who Made it*, Doha: Bloomsbury.

Spector, Ivar, 1958, 'Soviet cultural propaganda in the Near and Middle East', in Walter Z. Laqueur (ed.), *The Middle East in Transition*, New York: Praeger.

Stetkevych, Jaroslav, 2016, *The Hunt in Arabic Poetry*, Notre Dame: Notre Dame University Press.

Tamimi, Azzam, 2011, *Rached Ghannouchi: A Democrat within Islamism*, New York: Oxford University Press.

Tawasol, 2012, 'Tard Youssef Seddik min jamaʿa al-Zaytuna', *Tawasol TV*, 26 April, https://www.youtube.com/watch?v=-Zgp5MwM9QM (accessed 8 September 2017).

Tawfiq, Ahmed Khaled, 2008, *Yutubiya*, Doha: Bloomsbury Qatar Foundation.

—, 2011, 'Inihum yaʾkulun al-Kintaki', Aktowfik Blog, 3 February, http://aktowfik.blogspot.com/2011/02/blog-post_05.html (accessed 28 February 2018).

—, 2012, *al-Sinja*, Cairo: Dar al-Shorouk.

—, 2015, *Mithl Ikarus: Riwaya*, Cairo: Dar al-Shorouk.

—, 2017, *Al-Ghuz waraʾ al-sotur*, Cairo: Dar al-Shorouk.

Taylor, Adam, 2016, 'When Trump calls Obama the "founder of ISIS" he sounds like a Middle East conspiracy theorist', *The Washington Post*, 11 August, https://www.washingtonpost.com/news/worldviews/wp/2016/08/11/when-trump-calls-obama-the-founder-of-isis-he-sounds-like-a-middle-east-conspiracy-theorist/?utm_term=.0aae9965ce98 (accessed 12 August 2016).

Taylor, Philip M., 2007, *The Projection of Britain: British Overseas Publicity and Propaganda, 1919–1939*, Cambridge: Cambridge University Press.

The Telegraph, 2009, 'Video "shows Iranian woman bleeding to death in Tehran"', *The Telegraph*, 21 June, http://www.telegraph.co.uk/news/worldnews/middle east/iran/5595701/Video-shows-Iranian-woman-bleeding-to-death-in-Tehran. html (accessed 6 August 2017).

—, 2011, 'Die hard in Derna', *The Telegraph*, 31 January, http://www.telegraph. co.uk/news/wikileaks-files/libya-wikileaks/8294818/DIE-HARD-IN-DERNA. html (accessed 1 November 2017).

Le Temps, 2014, 'La HAICA face aux desperados', *Le Temps*, 15 October, http:// www.letemps.com.tn/article/86856/la-haica-face-aux-desperados (accessed 15 October 2014).

Thoraval, Yves, 1975, *Regards Sur Le Cinéma Egyptien*, Beyrouth: Darl el-Machreq.

Timberg, Craig, Elizabeth Dwoskin, Adam Entous and Karoun Demirjian, 2017, 'Russian ads, now publicly released, show sophistication of influence campaign', *The Washington Post*, 1 November, https://www.washingtonpost.com/business/ technology/russian-ads-now-publicly-released-show-sophistication-of-influe nce-campaign/2017/11/01/d26aead2-bf1b-11e7-8444-a0d4f04b89eb_story. html?utm_term=.a074951c9ccb (accessed 13 November 2017).

Tor, D. G., 2015. 'God's cleric: Al-Fuḍayl b. 'Iyāḍ and the transition from caliphal to prophetic sunna', in Behnam Sadeghi, Asad Q. Ahmed, Adam J. Silverstein and Robert G. Hoyland (eds), *Islamic Cultures, Islamic Contexts: Essays in Honor of Professor Patricia Crone*, Leiden: Brill.

Touzet, Jean Louis, 2013, 'La feuille de route d'Aqmi au Mali', *Libération*, 25 February, http://www.liberation.fr/monde/2013/02/25/la-feuille-de-route-d-aqmi-au-mali_884410 (accessed 2 May 2013).

Trager, Eric, 2011, 'After Tunisia, is Egypt next?', *The Atlantic*, http://www.theat lantic.com/international/archive/2011/01/after-tunisia-is-egypt-next/69656/ (accessed 19 October 2017).

Trofimov, Yaroslav, 2016, 'How Tunisia became a top source of Tunisian recruits', *The Wall Street Journal*, 25 February, https://www.wsj.com/articles/how-tunisia-became-a-top-source-of-isis-recruits-1456396203 (accessed 9 March 2018).

Troudi, Mounir, and Ammar404tounsi, 2010, *YouTube*, 25 May, https://www. youtube.com/watch?v=YRZipuMCxkI (accessed 6 February 2018).

Tunisia Live, 2012, 'The memories of journalist Sofiene Chourabi', *YouTube*, 18 January, https://www.youtube.com/watch?v=I8h5yv9BTf0 (accessed 15 February 2018).

Tunisie numerique, 2012, 'Abu Iyadh: L'illustre inconnu', *Tunisie numerique*, 18 September, http://www.tunisienumerique.com/abou-iyadh-lillustre-inconnu/144762 (accessed 13 March 2013).

Al-Tunisiya, 2014, 'Abu Loqman yakhlufu Abu 'Ayadh 'ala ra's Ansar al-Shari'a fī Tunis', *Al-Tunisiya*, 27 February, http://www.attounissia.com.tn/details_article.php?t=42&a=114779 (accessed 1 May 2013).

Turck, B., 1972, 'The authoritative Al-Ahram', *Aramco World* 23.5, http://archive.aramcoworld.com/issue/197205/the.authoritative.al-ahram.htm.

UNICEF Statistics, 2017, https://www.unicef.org/infobycountry/egypt_statistics.html#117 (accessed 28 March 2017).

United Nations Human Settlements Programme, 2012, *The State of Arab Cities 2012: Challenges of Urban Transition*, http://www.unhabitat.org/pmss/listItemDetails.aspx?publicationID=332 (accessed 1 May 2013).

United Nations Security Council, 2017, *Consolidated United Nations Security Council Sanctions List*, https://www.un.org/sc/suborg/en/sanctions/un-sc-consolidated-list (accessed 26 February 2018).

US Department of Justice, 2018, 'Internet Research Agency indictment', https://www.justice.gov/file/1035477/download (accessed 10 July 2018).

van der Linden, Sander, Anthony Leiserowitz, Seth Rosenthal and Edward Maibach, 2017, 'Inoculating the public against misinformation about climate change', *Global Challenges*, 1.2.

Vatikiotis, P. J., 1971, 'The corruption of futūwa: a consideration of despair in Nagib Maḥfūẓ's Awlād Ḥāratinā', *Middle Eastern Studies* 7.2, 169–84.

—, 1978, *Nasser and His Generation*, New York: St Martin's Press.

Voice Of America, 2014, 'Tribal leaders unveil "Save Libya" plan', *VOA*, 13 January, https://www.voanews.com/a/tribal-leaders-unveil-save-libya-plan/1829342.html (accessed 22 January 2018).

Wahl, Jean, 1926, *Étude sur le Parménide de Platon*, Paris: F. Rieder & Cie.

—, 2005, *Les philosophies pluralistes d'Angleterre et d'Amérique*, Paris: Les empêcheurs de penser en rond.

Al-Watan, 2014, 'Fadiha Fairmont: Safqa ma'a al-shaytan', *Al-Watan*, 17 February, http://www.elwatannews.com/news/details/419569 (accessed 21 July 2017).

The White House, 2011, 'Remarks by the President on the Middle East and North Africa', The White House Archives, 19 May, https://obamawhitehouse.archives.gov/the-press-office/2011/05/19/remarks-president-middle-east-and-north-africa (accessed 28 February 2018).

Williams, Raymond, 1976, *Communications*, New York: Penguin Books.

—, 1977, *Marxism and Literature*, Oxford: Oxford University Press.

—, 1986, 'The uses of cultural theory', *The New Left Review* I.158, https://new leftreview.org/I/158/raymond-williams-the-uses-of-cultural-theory (accessed 28 February 2018).

Wolf, Anne, 2017, *Political Islam in Tunisia*, Oxford: Oxford University Press.

World Bank, 2017a, 'Individuals using the Internet (percent of population)', *WorldBank*, https://data.worldbank.org/indicator/IT.NET.USER.ZS?end= 2015&locations=EG&start=1999 (accessed 3 June 2017).

—, 2017b, 'Mobile cellular subscriptions (per 100 people)', *WorldBank*, http://data. worldbank.org/indicator/IT.CEL.SETS.P2?end=2015&locations=EG&start= 1999 (accessed 3 June 2017).

Wright, Dean, 2009, 'The fall of the wall – and the media's role', *Reuters*, 9 November, http://blogs.reuters.com/fulldisclosure/2009/11/09/the-fall-of-the-wall-and-the-medias-role/ (accessed 30 July 2017).

Youm 7, 2017, 'Bi-maʿradh al-kitab . . .', *Youm 7*, 24 January, http://www.youm7. com/story/2017/1/24/%D8%A8%D9%85%D8%B9%D8%B1%D8%B6-%D8%A7%D9%84%D9%83%D8%AA%D8%A7%D8%A8-%D8%B3 D9%84%D8%B3%D9%84%D8%A9-%D8%A7%D9%84%D8%A7% D8%AD%D8%AA%D9%84%D8%A7%D9%84-%D8%A7%D9%84% D9%85%D8%AF%D9%86%D9%89-%D9%84%D9%80-%D8%B9% D9%85%D8%B1%D9%88-%D8%B9%D9%85%D8%A7%D8%B1-% D8%B9%D9%86-%D8%AF%D8%A7%D8%B1/3070350 (accessed 8 January 2018).

Young, Michael, 2007, *Death, Sex & Money: Life Inside a Newspaper*, Carlton: Melbourne University Press.

Yusuf, Walid, 2012, 'Al-Liqa' al-sahafi maʿa Abu ʿAyadh al-Tunisi', *Muslim.org*, 1 January, http://www muslm.org/vb/archive/index.php/t-465707.html (accessed 1 October 2012).

Zaky, Amir, 2012, 'When Dracula speaks', *Egypt Independent*, 11 August, http:// www.egyptindependent.com/news/when-dracula-speaks-arabic (accessed 27 February 2018).

Zayani, Mohamed, 2015, *Networked Publics and Digital Contention: The Politics of Everyday Life in Tunisia*, Oxford: Oxford University Press.

Zayit, Kamal, 2014, 'Al-jamaʿat al-musallaha fi Jaza'ir min "Buyaʿli" ila ʿjund al-khilafa"', *Al-Quds at-Arabi*, 11 October, http://www.alquds.co.uk/?p=233522 (accessed 2 June 2017).

Zeilin, Aaron, 2012, 'Know your Ansar al-Sharia', *Foreign Policy*, 21 September, http://foreignpolicy.com/2012/09/21/know-your-ansar-al-sharia/ (accessed 17 January 2018).

Zemni, Sami, 2013, 'From socio-economic protest to national revolt: the labor origins of the Tunisian Revolution', in Nouri Gana (ed.), *The Making of the Tunisian Revolution*, Edinburgh: Edinburgh University Press.

Index

EU representative:
Easy Access System Europe
Mustamäe tee 50, 10621 Tallinn, Estonia
Gpsr.requests@easproject.com

www.ingramcontent.com/pod-product-compliance
Lightning Source LLC
Chambersburg PA
CBHW050633280326
41932CB00015B/2624